THE WEATHER HANDBOOK

THE WEATHER HANDBOOK

MARIA COSTANTINO

Published by SILVERDALE BOOKS
An imprint of Bookmart Ltd
Registered number 2372865
Trading as Bookmart Ltd
Blaby Road
Wigston
Leicester LE18 4SE

D&S Books Ltd
Kerswell,
Parkham Ash, Bideford
Devon, England
EX39 5PR

e-mail us at:- enquiries@d-sbooks.co.uk

This edition printed 2005

ISBN 1-84509-187-6

DS0129. Weather Handbook

Creative Director: Sarah King
Editor: Nicky Barber
Project editor: Anna Southgate
Designer: Big Metal Fish

Fonts used within this book: Abadi, Goudy and Helvetica

Printed in Thailand

1 3 5 7 9 10 8 6 4 2

CONTENTS

INTRODUCTION

WEATHER AND CLIMATE

ALTHOUGH INTERRELATED, there is a great difference between the weather and climate. The weather is the state of the atmosphere at a particular time and in a particular place – the sunshine we have today, and the rain we had yesterday. The climate is the weather over a long period. Climates have patterns that can be examined year after year to produce information about averages – annual average rainfall, average hours of sunshine, average temperatures – as well as to see if there are patterns in extreme variations from these averages.

The climate has been responsible for the development of the worlds' civilisations throughout the ages: long-term, unchanging (or at least imperceptibly changing) climatic conditions encouraged the development of farming, towns, cities and cultures. Changes in climate such as long Ice Ages followed by thaws and flooding, or the encroaching sands of the desert, have in the past caused the downfall of major cultures.

Weather – particularly in its extreme forms – has also played its part in shaping history. In 56BC many of Julius Caesar's ships were destroyed by a storm in the English Channel, forcing him to delay his planned invasion of Britain. When the Mongol leader Kublai Khan, grandson of the mighty Genghis Khan, attempted to invade Japan in 1274 he saw his fleet of 700 ships overwhelmed by a typhoon with the loss of more than 10,000 lives. Khan tried his luck again in 1281, but another typhoon struck and his dreams of empire were sunk, along with his fleet. When King Philip II of Spain sent the Armada to invade England in 1588, it wasn't the superiority of the English fleet that brought grief to the Spanish but an unforeseen gale in the English Channel. And in 1776, American

RIGHT: *Encroaching deserts today threaten large areas of the world and its population.*

ABOVE: *Even today, the weather can have disastrous effects – shipwrecks are common throughout the world.*

general George Washington's revolutionary forces were trapped and surrounded in a sea of mud at Valley Forge. But a north wind brought lower temperatures and the mud froze, allowing the American troops to escape the British forces, and the rest, as they say, is history!

Humans have tried for thousands of years to predict the weather. Our quest to answer the seemingly simple question 'What will the weather be like tomorrow?' started in the Stone Age and used the same, fundamental technique that we use today: observation.

ABOVE: *Methods of forecasting the weather have been practised since the early days of mankind.*

Before the end of the last Ice Age, around 8500BC, the sea level was around 120 m (390 ft) lower than it is today. There were land bridges between France and England, between Siberia and Alaska, across the Bosphorus, and between Australia and New Guinea, as well as 'bridges' connecting many of the Indonesian islands with the Malayan peninsula. These land bridges were important migration routes for many land-dwelling animals, and for humans. By around 7500BC, virtually all of these land bridges had disappeared under the melting waters as temperatures began to climb. Ever-adaptable and ever-resourceful, despite these major changes, humans continued to thrive.

BELOW: *The shapes of landmasses – continents, archipelagos and individual islands – are, in part, due to the weather.*

THE HISTORY OF WEATHER

As HUNTER-GATHERERS, early humans used the weather – wind, rain and snow (which revealed the tracks of their prey) – and the seasonal migrations of animal herds to stalk their quarry. When humans turned to farming, around 4000BC, understanding the weather was vital. Good weather ensured a full crop and full bellies; bad weather could mean famine. The farmers of the Nile Valley of Ancient Egypt may have visualised the river as the goddess Isis, but they clearly understood that their lives and livelihood depended on the annual flooding of the Nile to irrigate their fields.

Some of the earliest images made by humans include representations of the Sun, Moon and rain, although early weather 'forecasting' was not much more than divination – trying to predict the future by supernatural or intuitive means, rather than scientific or rational methods. One of the earliest systems for classifying and interpreting weather-related events was developed by the ancient Babylonians in the 12th century BC. The Babylonians believed that atmospheric phenomena depended on the movement of the Moon and stars. However, it is the Chinese who are credited with the first regular weather observations. Inscriptions dating from the Shang dynasty (1766-1122BC) reveal observations of wind directions, heights of snow, and the appearance of the sky on ten consecutive days. In 1066BC, during the Chou dynasty, the Chinese began the world's first official record-keeping of weather-related phenomena. These 'retrospective' views of weather events allowed the Chinese to build up a picture of climatic pattern.

ABOVE: *The Moon has an orbit of 27.3 days.*

In Europe, the first attempts to understand the weather began in ancient Greece. In 600BC Thales (c.625-c.546BC) devised a weather calendar for sailors and he is said to have predicted the solar eclipse of 585BC. According to legend, Thales bought a huge crop of olives shortly before a drought, cornered the market in a time of shortage (no doubt inflating the prices of his olives!) and became very rich. The Greek philosopher Aristotle (384-322BC) wrote Meteorologica, the first wide-ranging book on the weather, in which he discussed clouds, snow, rain, storms and hail, as well as 'optical phenomena' such as rainbows and halos. Although by contemporary standards Aristotle's work is flawed, he was the first to recognise that the difference between rain, snow and frost was to do with varying degrees of temperature.

The Ancient Romans continued the tradition of observing weather phenomena. The poet Virgil (70-

19BC) recorded his observations in his pastoral poem the Georgics, using the habits of birds and animals to guide him, while another poet, Lucretius (c.95-c.45BC), discussed lightning and water spouts in his book *De rerum natura*. Gaius Pinius Secundus, better known as Pliny the Elder (AD23-79), devoted his life to trying to understand the world around him by observing all aspects of nature, including the weather. His depth of interest and his quest for knowledge unfortunately led to his untimely death: seeking to observe at close quarters the eruption of Mount Vesuvius in AD79 (which famously overwhelmed the towns of Pompeii and Herculaneum), Pliny was suffocated by the toxic fumes of the volcano's pyroclastic blast.

Yet despite the advances made by the Greeks and Romans in this field, it was not until the 17th century that scientific analysis and experimentation began to uncover some of the mysteries of the natural world, including the weather.

ABOVE: *Pliny's quest to understand the world around him led to his death during the eruption of Vesuvius.*

Witchcraft and Weather Prediction

Until the 17th century, attempting to predict the weather was part of astrology, which was classed by the Christian church in the Middle Ages as an occult science. As far as most ordinary people were concerned, God made the weather, and he alone had the power to change it. People offered prayers to interceding saints for particular weather such as rain, or for protection from the elements. For example, it was widely believed that St Donatus had miraculously survived a lightning strike and therefore provided protection from lightning. If his protection was insufficient, then it was all the work of the Devil and his witches!

While predicting the weather was off-limits, there are still records of reported weather – usually extreme – although there were rarely any attempts to explain why it happened. We know there were some extremely cold winters in Europe: in 1011 the seas froze in the northern Adriatic and around Constantinople (Istanbul); in 1302 French writers reported that it was so cold in December that people were found frozen to death in their beds; while the period between 1313 and 1321 in western Europe saw catastrophic weather – cold, wet summers, and wet autumns followed by wet springs – resulting in failed harvest and famine during which sheep, cattle and humans perished in their thousands. Even the arrival of the Black Death (bubonic plague) in England in 1349 can now be traced back to a climatic catastrophe which occurred 17 years earlier: the worst flood in the first millennium not only drowned 7 million people in China, it also wiped out the natural habitat of the black rat which began its westward migration across Asia into Europe.

Despite the passing of laws in England in 1677,

LEFT: *Weather prediction was considered an occult science in Europe until the 17th century.*

LEFT: *Ice storms are rare, but those on record have been devastating.*

MAIN IMAGE: *Making observations of the visible aspects of the weather – the colour of the sky and clouds – or of the seasonal habits of animals, led to the development of weather 'lore'.*

which stated that all 'rainmakers and weather seers' would be burned at the stake (a law that appears to have been largely ignored and its enforcement overlooked as it wasn't actually repealed until 1959!), a great deal of knowledge about the weather was accumulated, largely by country folk, the people whose lives and livelihoods depended on it. The weather 'lore' which still exists today in popular sayings such as 'Red sky at night, shepherd's (or sailor's) delight; red sky in the morning, gives shepherds (and sailors) a warning' or 'Rain before seven, Fine by eleven' was based on careful observations and has some meteorological value.

More sayings:

'A ring around the Sun or Moon,
Means that rain is coming very soon.'

'If November ice will bear a duck,
There'll be nothing after but slush and muck!'

'He who shears his sheep before St Saviour's Day*
Loves his wool more than his sheep.'

(St Saviour's Day falls on 13 May, St Mamertus's Day on 11 May and St Pancras's Day on 12 May: these dates coincide with the ancient (pagan) festival of the 'Ice Saints'.)

'St Swithin's Day*, if ye do rain,
For forty days it will remain;
St Swithin's Day and ye be fair,
For forty days twill rain no more.'

(St Swithin's day is 15 July: it marks the interment at Winchester Cathedral of Swithin, the city's archbishop in the 9th century.)

The Age of Science

While country folk were scanning the Earth and the skies for signs of weather, such as shooting stars, which many believed foretold a severe winter, or comets that presaged strong winds and droughts, by the 17th century, rational thought was challenging the established beliefs of the Church. This would lead to a range of major scientific discoveries. Some of the early investigators of meteorological phenomena were scientists, chiefly astronomers and mathematicians, but there were also sea captains, architects, priests and teachers – as well as a kite-flying diplomat – among their ranks!

One of the first radical challenges to accepted belief was made by the Italian astronomer Galileo Galilei (1564-1642). Even when being examined by the Catholic Church's Inquisition, which forced him to deny that the Earth revolves around the Sun, he stated '*E pur si muove!*' ('Yet it does move!'). Galileo's heliocentric (Sun-centred) view of the Solar System would later have a far-reaching impact on human understanding of the nature of the atmosphere.

While Galileo also devised some primitive equipment for observing fluctuations in temperature and atmospheric pressure, it was his student Evangelista Torricelli (1608-1647) who, in 1643, demonstrated the existence of air pressure and invented the barometer to measure it. Torricelli showed that the pressure air exerts on the Earth's surface supports a column of mercury 76 cm (29.92 in) in height.

ABOVE: *Galileo Galilei developed the heliocentric (sun-centred view of the Solar System in which the Earth – and other planets – orbit the Sun.*

MEASURING GAUGES AND SYSTEMS

The first sealed, liquid-in-glass air thermometer, based on principles set out in 1597 by Galileo, was developed in 1641 at the court of the Grand Duke Ferdinand of Tuscany. But the problem of 'stem calibration' – what unit of measurement to use to mark graduated scales – was fiercely debated. It was in 1694 that Carlo Renaldini, a physicist from Padua, proposed that the freezing and boiling points of water be used as fixed points on thermometric scales. In 1714, the German physicist Daniel Fahrenheit invented the thermometer named after him and was the first to measure mercury temperature. In 1742 Andres Celsius introduced the centigrade scale.

Devices for measuring rainfall had been around for some time. The architect of St. Paul's Cathedral in London, Sir Christopher Wren designed a rain gauge as well as a 'weather clock'. Invented in 1663, Wren's device was one of the first self-registering instruments for measuring weather change. Driven by a clock, it made pencilled 12-hour readings of a barometer and a wind vane.

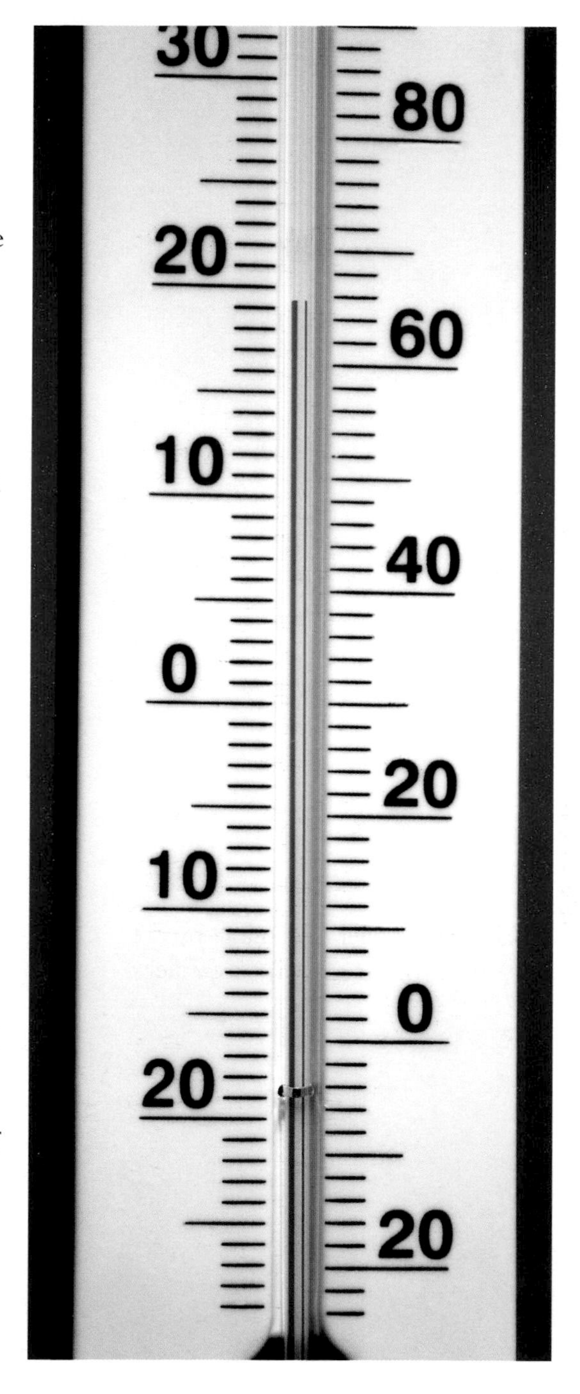

The first effective instrument for measuring humidity, called a hygroscope, was described by the German cardinal Nicolaus de Cusa at the end of the 15th century. The hygroscope was simply a balance or set of scales: on one side were stones, on the other side the equivalent weight in dry wool. In dry air, the two sides balanced, but when the air was damp this affected the weight of the wool, and tipped the balance. A similar device, exploiting the water-retaining qualities of the bristles of an ear of wild oat, was invented by Robert Hooke in England in the 17th century. But while these devices demonstrated that there was moisture in the air, they couldn't accurately show how much moisture there was. Around 1780, the first mechanical 'hygrometer' was built by the Swiss physicist and geologist Horace Benedict de Saussure. He used a human hair – which stretches when damp – to measure humidity.

NEWTON, BOYLE AND THE SECOND LAW OF MOTION

At the same time that Galileo was experimenting in Italy, the Danish astronomer Tycho Brahe, the English priest Jeremiah Horrocks, and the German astronomer Johannes Kepler were undertaking studies that had no immediate connection with the weather but would in the future play a vital role in developing the science of meteorology. They were all studying the behaviour of the Solar System and the nature of the laws that governed it. Brahe, Galileo and Horrocks plotted the distances between the Earth and the Sun, and the Earth and the Moon, and, in 1609, Kepler showed that the Earth's orbit around the Sun was elliptical.

Kepler's work in particular was to be of great importance to Sir Isaac Newton, whose *Philosophiae Naturalis Principia Mathematica* (1687) is generally held to be one of the most outstanding achievements in the history of science. In it, Newton argued that all matter – from the tiniest particle to the largest planet – responded to certain laws of gravitation and motion. The Second Law of Motion is one of the fundamental points of meteorology. It says that a body whose motion is changed by an outside force will accelerate in the same direction as that force and at a rate that is in direct proportion to the amount of the force. It is expressed mathematically as $F=ma$ (force equals mass times acceleration). However, although force, mass and acceleration are major factors in the Earth's atmosphere, this law still didn't explain the workings of the weather!

The Irish physicist Robert Boyle, contemporary of Newton, worked out the relationship between the volume of a body of air and its pressure. Boyle's Law states that 'at a constant temperature, the volume of a gas varies inversely with its pressure'. In other words, decrease the size of a body of air, and you increase the pressure, and vice versa.

ABOVE: *Newton's second law of motion is one of the lynchpins of meteorology.*

UP, UP AND AWAY

In the 18th century, French physicist Jacques Charles worked out the connection between the volume of gas and temperature. Charles's Law states that if the pressure of a gas remains constant, but its volume changes, there is an accompanying proportional change in temperature. In other words, the hotter a gas, the bigger its volume. The fundamental law of atmospheric dynamics – hot air rises because it's lighter than the cooler air around it – also allowed Charles, in December 1783, to make one of the first ever meteorological flights in a balloon. By taking a barometer with him, Charles was able to work out his altitude by measuring the drop in air pressure as his balloon gained height over Paris.

Balloons were soon to become a feature of meteorological science. British scientist James Glaisher made his first balloon flight in 1862 with an instrument panel consisting of several thermometers, two barometers, a compass and, oddly, a pair of opera glasses. At 8,850 m (29,000 ft) Glaisher fainted from lack of oxygen. Fortunately, his pilot was made of sterner stuff and though nearly rigid with cold, managed to open a control valve on the balloon with his teeth to bring them both safely back to Earth!

A safer balloon method was pioneered by Leon Teisserenc de Bort, founder of the Observatory of Dynamic Meteorology near Paris. Unmanned high altitude observations were made with his instrument-bearing balloons and kites. Some were anchored to the ground, others were 'free-flying' which made recovering the recording instruments and the information a bit tricky. Things were made somewhat easier in the 1930s when Robert Bureau and Pavel Malachanov developed the technique of radiosonde observation. They placed radio transmitters into weather balloons so that the readings could be radioed back to ground stations. A radiosonde network evolved in the 1940s to monitor and transmit data about upper-air humidity, pressure and temperature.

Today, there are various forms of meteorological balloon. Some are designed to remain at a constant height and for ultra-long periods. The simplest is the un-instrumented pilot balloon, usually inflated with hydrogen, which is used to determine cloud heights. Similar balloons fitted with radar reflectors are designed to examine high-altitude winds. The largest balloons, inflated with helium, are for observations of the Sun and cosmic rays, while the aptly named 'rockoon' is system-designed to investigate the upper atmosphere. A research rocket is lifted by balloon into the stratosphere (see page 54) and then fired to examine the arrival of energy

ABOVE: *A pilot balloon is a simple meteorological balloon without instruments, and is used to determine cloud heights.*

LEFT: *John Dalton, whose law of partial pressures allows meteorologists to calculate the amount of water vapour in the air, and thereby understand how clouds are formed.*

Dalton's Law

Between 1788 and 1792, English chemist John Dalton undertook numerous experiments regarding barometric pressure. Dalton was intrigued by the behaviour of the barometer in certain atmospheric conditions, the directions of winds and the heating and cooling of air. The outcome of his experiments, sometimes called Dalton's Law, sometimes the Law of Partial Pressures, was the understanding that in a mixture of gases, each gas exerts the same pressure as it would if it were on its own. It also says that the total pressure of the mixture of gases is equal to the sum of their individual pressures. Dalton's Law allows meteorologists to calculate the amount of water vapour in the air and thereby understand cloud formation and precipitation – fog, rain and snow – in mathematical terms.

THE FIRST METEOROLOGICAL MAP

In the 17th century, as great advances in science and understanding the natural world were made, some people began to apply these newly discovered laws to the study of the weather. Chief among these was the British Astronomer-Royal, Sir Edmund Halley, who discovered the comet that bears his name. Halley's numerous contributions to the science of the weather include his publication in 1686 on the causes of the tropical trade winds and monsoons, which he illustrated this memoir with history's first meteorological map. Halley's global wind circulation chart is based on his and other seafarers' observations made at sea and during a two-year stay on the island of St Helena off the west coast of Africa. Halley deduced that the trade winds and monsoons were caused by thermal convection – the actions of the Sun on the air over the Equator – which caused the air to rise and distribute outwards. But Halley couldn't explain why changes in the wind were directly linked to changes in barometric pressure or why the trade winds blow from the northeast in the northern hemisphere and from the southeast in the southern hemisphere.

HADLEY CELLS

Around 50 years later, in 1735, a London lawyer called George Hadley gave the world a much clearer picture of the trade winds. Thermal convection and solar radiation accounted for them, but Hadley recognised that the westward movement of the winds was caused by the west-to-east rotation of the Earth. In Hadley's model, air rises above the zone that is most strongly heated (the Tropics), flows at height towards the poles where its sinks, and completes its Equator-ward flow at low level. While modern theories

ABOVE: *Although best known for his observations of the comet named after him, Edmund Halley also contributed to the science of meteorology.*

suggest that a more horizontal flow plays a greater part in the general circulation, the concept does correspond approximately to the circulation in the troposphere (see page 52) between the Equator and 30°N and 30°S, so the term 'Hadley cell' is often applied to this circulation or 'loop'.

A STATESMAN IN A STORM

Benjamin Franklin was a printer, publisher, author, the first US Ambassador to France (1776-85), and instrumental in helping to draft the American Declaration of Independence and the US Constitution. He was also fascinated by violent storms. Through his famous, though hazardous, kite experiment of 1752, Franklin proved the electrical nature of lightning and subsequently invented the lightning conductor. From his experiments, Franklin formulated ideas about the role that electricity played in causing rain, about moist, heated air causing waterspouts, whirlwinds, thunderstorms, and about the Equator-to-Pole atmospheric circulation.

In 1743, while trying to observe an eclipse of the Moon in Philadelphia, Franklin discovered by accident the prevailing southwest to northeast drift of Atlantic coastal storms, despite the fact that they had strong winds in other directions. In Philadelphia, storm clouds blowing in from the northeast obscured his view of the Moon, but the eclipse was viewed in Boston as scheduled. Four hours later, the storm arrived in Boston. Using newspaper reports from New England to Georgia, Franklin concluded that the storms with winds from the northeast started out in the southwest – in Georgia and the Carolinas – and travelled towards the northeastern states at a rate of 160 km/h (100 mph). Franklin deduced that the reason for the northeast storms was 'some great heat and rarefaction in the air' over the Gulf of Mexico, while the strike of the storms on the coast and the ridge of the Appalachian Mountains guided the northeast flow of air. Franklin then went on to make an extensive study of the Gulf Stream –measuring its temperature and direction and estimating its speed and possible effects on the weather. He subsequently produced the first true chart of an ocean current.

MAIN IMAGE: *An image representing Franklin's 1752 experiment with a key and a kite to prove the electrical nature of lightning. Franklin's hazardous experiments led him to invent the lightning conductor.*

Meteorology Comes of Age

The new technologies of communication of the mid 19th century, such as the electric telegraph, played a great role in meteorological research – the weather could be observed, the findings transmitted and analysed all in a short space of time. However, it was an incident of extreme weather that galvanised nations into developing weather services. During the Crimean War, on 14 November 1854, the French warship Henri IV and 38 merchant ships were sunk in a 'sudden' fierce storm near Balaclava with the loss of four hundred lives. The French astronomer Urbain le Verrier was charged by the French government to investigate the disaster. Verrier found that the 'sudden' storm had in fact already formed two days earlier and had swept across Europe from northwest to southeast. Realising that storms travelled across the Earth's surface, Verrier suggested to the government that cities could exchange information on the weather, and use the information to locate storms and predict their movement.

ABOVE: *Robert FitzRoy*

In 1855, the French government set up 24 observation stations, 13 of which were linked by telegraph. Other European countries followed suit. In 1854, Admiral Robert FitzRoy was appointed superintendent of the newly established Meteorological Department of the Board of Trade which gathered information from over 40 weather stations around the British Isles and neighbouring continental Europe. In 1861, FitzRoy initiated the storm warning service for shipping (which

LEFT: *The eye of a storm seen from a satellite.*

continues today) and in 1864 began publishing general weather forecasts in *The Times*. Unfortunately, FitzRoy's forecasts were not always very accurate and the editor of *The Times* dropped them, shortly before FitzRoy committed suicide. In the USA, a weather service began in 1870 under the auspices of the Army Signal Corps which boasted a 75% accuracy rate in their predictions of rainfall and temperature.

SYNOPTIC METEOROLOGY

The claims of the Signal Corps to accuracy were, however, dubious. This is because, at the time, meteorologists were all relying on what is known as synoptic meteorology: simultaneous observations of the weather over a large area to make predictions of the weather's future movements. These observations were land-based and did not take into account topography – the 'shape' of the landscape. Nor did it take any account of the fact that clouds are steered by winds in the upper atmosphere. In order to predict cloud movement it was necessary to get information from the upper atmosphere, but, as we have seen, balloon flights were limited in their capabilities until the development of radiosonde in the 1940s. Fortunately, the development of air travel in the 20th century meant that aircraft could carry on-board equipment to measure temperature, humidity, wind and air pressure which could then be radioed back to 'met offices' on the ground.

The past 50 or so years have seen tremendous leaps in technology including supersonic jets, satellites, computers and space stations. This means that we now have greater knowledge of the previously 'invisible' thin, protective layer of gases that surrounds the Earth – the atmosphere – and its role in creating the weather. Understanding the atmosphere has given meteorologists the ability to forecast the weather with some accuracy up to six days ahead. Sometimes meteorologists still get the forecasts wrong – but at least they don't get burned at the stake these days!

LEFT: *Even with advanced technology, forecasters can still get it wrong!*

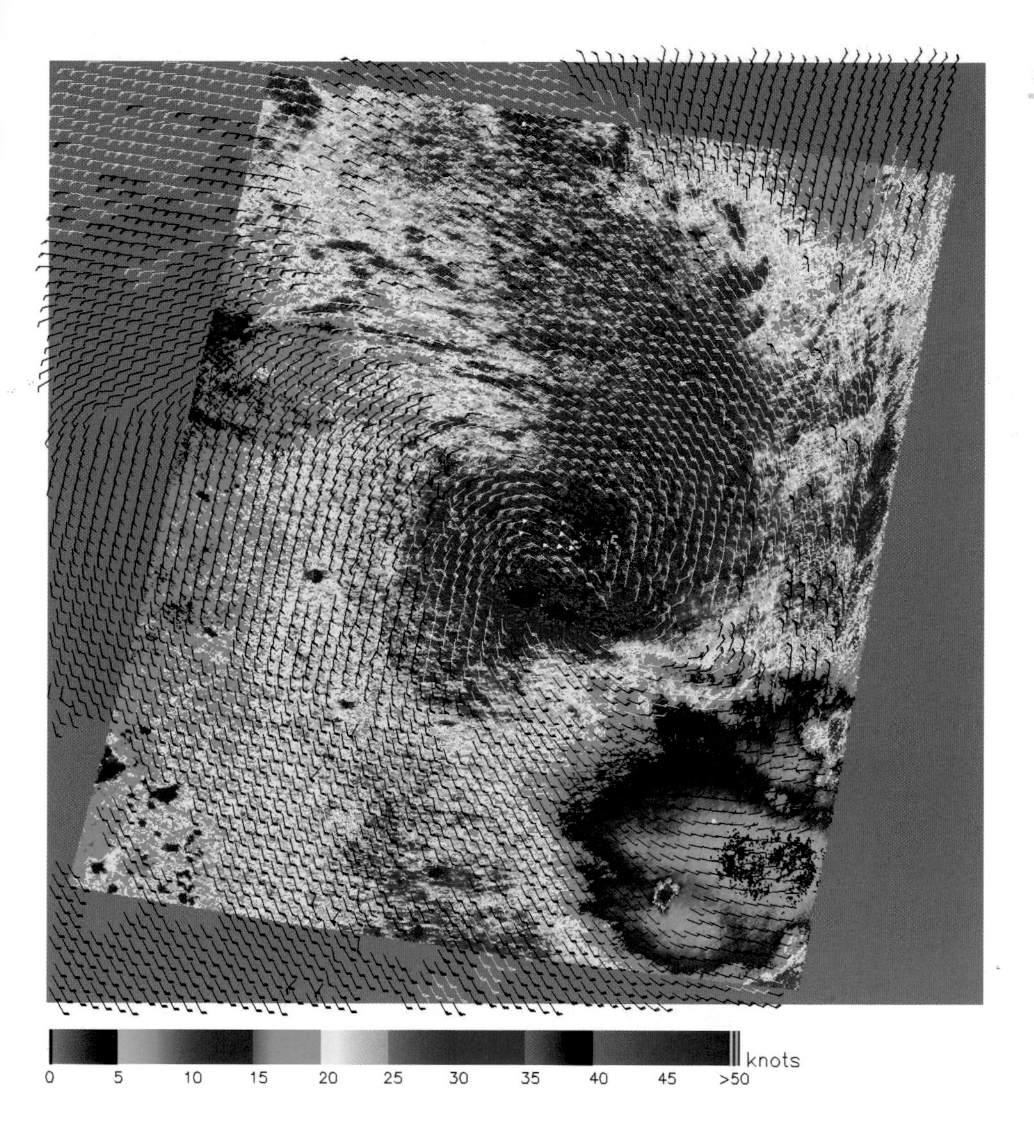

ABOVE: *The SeaWinds scatterometer aboard NASA's QuikSCAT satellite collected the data used to create this colourful image of Cyclone Olaf churning in the South Pacific on February 16, 2005. The coloured background shows the near-surface wind speeds at 2.5-km (1.5-mile) resolution. The strongest winds, shown in purple, are at the centre of the storm, with gradually weakening winds forming rings around the centre. The black barbs indicate wind speed and direction at QuikSCAT's nominal, 25-km (1.5 mile) resolution; white barbs indicate areas of heavy rain. NASA's Quick Scatterometer (QuikSCAT) spacecraft was launched from Vandenberg Air Force Base, California on June 19, 1999. QuikScat carries the SeaWinds scatterometer, a specialised microwave radar that measures near-surface wind speed and direction under all weather and cloud conditions over the Earth's oceans.*

THE WEATHER-MAKERS

THE SUN, THE EARTH AND THE ATMOSPHERE

The weather and the climates we experience on Earth are the products of the interaction between the Sun, the Earth, and the Earth's atmosphere.

THE SUN

OF ALL THE CELESTIAL BODIES, the Sun has the most profound influence on the Earth. Planet Earth receives just two-billionths of all the available solar radiation – a miniscule amount as far as the Sun is concerned – yet this is the source of all our light and heat. If it wasn't for that tiny amount of radiant energy, the weather – and life on Earth as we know it – would cease, and the temperature on the surface of the Earth would be no higher than -250°C (-418°F).

The Sun, just one of the hundred billion stars in our galaxy, is an enormous ball of hot gas about 1.4 million km (865,000 miles) in diameter and contains within it about 99% of the mass of our Solar System. Located some 150 million km (93 million miles) from Earth, the Sun is 1,306,000 times bigger than Earth: if the Universe was shrunk so the Earth was about the size and weight of a tennis ball, the Sun would measure about 4 m (12.5 ft) in diameter and weigh around 3 tonnes (3.5 tons).

RIGHT: *The Sun, source of Earth's light and heat.*

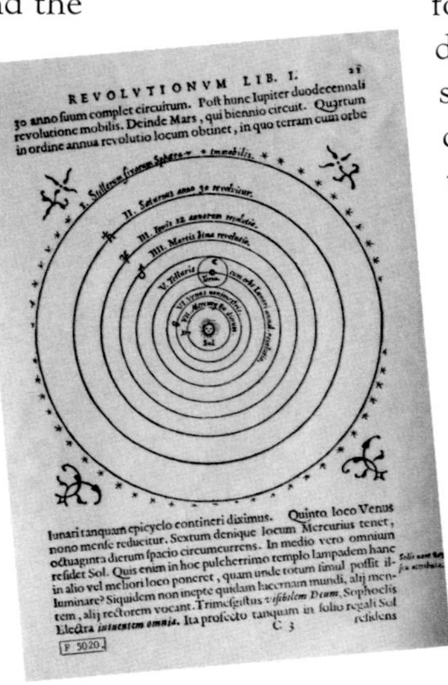

REVOLVTIONVM LIB. I. 21

30 anno suum complet circuitum. Post hunc Iupiter duodecennali revolutione mobilis. Deinde Mars, qui biennio circuit. Quartum in ordine annua revolutio locum obtinet, in quo terram cum orbe lunari tanquam epicyclo contineri diximus. Quinto loco Venus nono mense reducitur. Sextum denique locum Mercurius tenet, octuaginta dierum spacio circumcurrens. In medio vero omnium residet Sol. Quis enim in hoc pulcherrimo templo lampadem hanc in alio vel meliori loco poneret, quam unde totum simul possit illuminare? Siquidem non inepte quidam lucernam mundi, alij mentem, alij rectorem vocant. Trimegistus *visibilem Deum*, Sophoclis Electra *intuentem omnia*. Ita profecto tanquam in solio regali Sol residens

C 3

F 5020

ABOVE: *Copernicus produced the first scientifically convincing heliocentric theory.*

THE EARTH'S ORBIT

The Polish astronomer Nicolaus Copernicus (1473-1543) is famous for providing the first detailed and scientifically convincing argument that the Earth and the planets orbit the Sun, and that the Earth spins on its own axis. This heliocentric theory of the Universe was opposed to the prevailing views of the time –that the Earth was stationary and at the centre of the Universe – and would take more than a century to be accepted. We now know that the Earth orbits the Sun once every

365.25 days in an elliptical orbit, while at the same time, the Earth rotates on its axis in 24 hours, so points on the Earth's surface pass from full sunlight into shadow and back, causing day and night.

The different seasons we experience are the result of the Earth's axis throughout its elliptical orbit around the Sun: the Earth is 'tilted' at 23.5° from vertical. This means that, at the summer solstice in June, the northern hemisphere points towards the Sun and the north polar region is bathed in sunlight all day. At the south polar region, however, there is darkness until the situation is reversed at the winter solstice in December, when the North Pole experiences its winter night-time. The tilt of the Earth is also responsible for determining the extent of the Tropics: the Sun is

BELOW: *The seasons are caused by Earth's tilted axis and its elliptical orbit around the Sun.*

overhead at midday at the Tropic of Cancer (23.5°N) in June, and overhead at midday at the Tropic of Capricorn (23.5°S) in December.

A peculiarity of the elliptical orbit also means that, as winter approaches in the northern hemisphere, the Earth swings closer to the Sun reaching the perihelion (the closest point to the Sun) on 2 January. On that day Earth is separated from the Sun by a mere 147 million km (90 million miles) and the amount of solar radiation intercepted on Earth is at its highest. Consequently, the Earth receives more light and heat in the northern hemisphere's winter than it does in summer – although it doesn't seem that way! This is because in winter, the Sun's rays strike their part of the Earth at an angle that spreads the energy over a larger area, making the heat and light less concentrated. The Earth and the Sun are furthest apart at the aphelion, on 5 July when the distance between the two is 152 million km (95 million miles).

THE SUN AND ITS ENERGY

The Sun is one mighty thermonuclear powerhouse: temperatures at the core of this fiery ball are estimated to be as high as 15 million°C (27 million°F), while surface temperatures range between 4,000° and 6,000°C (7,200-10,800°F). Nuclear fusion at the Sun's core converts hydrogen to helium in a

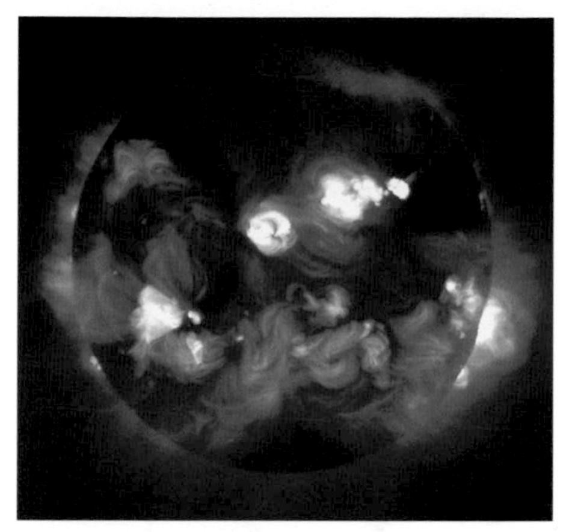

ABOVE: *Surface temperatures of the Sun can reach 6,000°C (10,800°F).*

process that creates huge amounts of energy – about the same as burning 500 million tonnes (550 million tons) of oil per second – which moves outwards from the core by radiation and convection until it reaches the photosphere, the Sun's visible surface, and escapes as radiation.

Above the Sun's photosphere are two outer layers: the chromosphere and the corona. The corona emits a continuous flow of electrically charged particles known as the solar wind, which stream out across the Solar System. The solar wind is modified by bursts of activity around the Sun's surface and atmosphere: these include solar flares and Coronal Mass Ejections (CMEs). When the energised particles from the solar wind, flares and CMEs reach Earth, most are deflected by the magnetosphere (the magnetic field) which surrounds the Earth, but

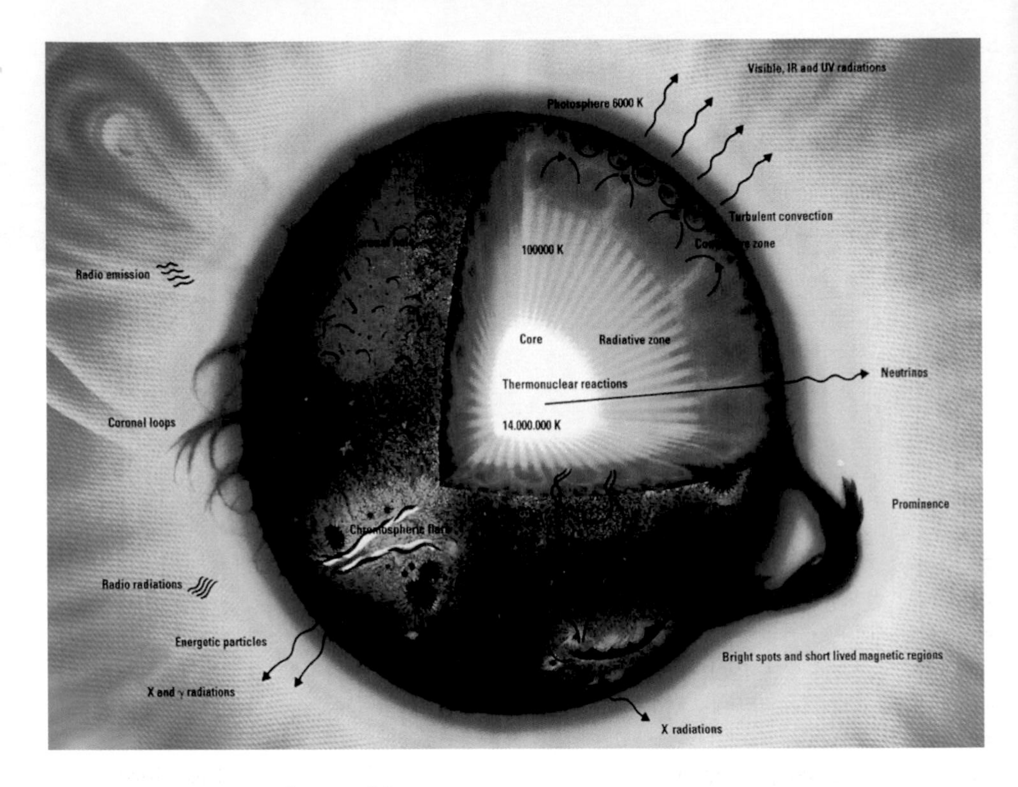

ABOVE: *Temperatures at the core of the Sun are estimated to be as high as 15 million°C (27 million°F). At the core, nuclear fusion converts hydrogen to helium in a process that creates a huge amount of energy which radiates from the core to the photosphere – the visible surface of the sun – where temperatures range from 4,000–6,000°C (7,200°F–10,800°F). Above the photosphere are two further layers: the chromosphere and the corona. The corona emits electrically charged particles known as solar wind, which are deflected by the Earth's magnetic field (the magnetosphere).*

some are channelled into the Earth's atmosphere above the magnetic poles. At the poles, the particles interact with gases in the atmosphere and the results are often visible in the beautiful shimmering light effects called aurorae which occur at altitudes of 90-300 km (56-187 miles). When a large burst of particles hits the magnetosphere, the result can be a magnetic storm that causes interference with electrical systems (such as radio and television transmissions) on Earth.

THE SOLAR CONSTANT

Of the enormous amount of energy that the Sun produces – 70,000 horsepower for every square metre of its surface – Earth receives only a minute fraction, about one two-billionth. This tiny proportion, equal to about 23 billion horsepower, is more energy every minute than humans use in all forms in one year. The Sun continuously bathes the outer reaches of the Earth's atmosphere in a radiant energy which has an average value of 1,370 watts/metre2. This is called the Solar Constant and it is this energy that drives the atmosphere and global weather. If the Sun's heat dropped by 13%, it is estimated that the Earth would be encased in a layer of ice 2 km (1 mile) thick; if its heat increased by 30%, all life on Earth would be cooked!

The Sun beams its energy through space to Earth as three forms of electromagnetic wave. Electromagnetic waves are similar to radio waves, but are much shorter in length, and they are measured by the distance between their peaks or crests.

The first electromagnetic waves are ultraviolet rays: these measure from 400 thousand-millionths to about 16 millionths of an inch. The second are infrared rays, measuring from 30 millionths to 400 thousands of an inch. Neither of these two wavelengths can be perceived by the human eye, although some animals, especially birds and bees, can see infrared rays. In between the ultraviolet and infrared rays are the wavelengths of the visible spectrum, the colours of the rainbow which, when mixed, produce white light. The wavelengths of the visible spectrum range from 16 millionths of an inch (violet) to 30 millionths of an inch (red). By comparison, radio wavelengths can measure several feet in length.

THE SOLAR BUDGET

The Earth absorbs radiant energy from the Sun, is warmed by it and re-radiates this energy, in the form of heat, into the atmosphere. The balance between the amount of solar energy that is received on the Earth's surface, in its atmosphere or in clouds, and the amount of energy that is reflected or radiated back into space, is known as the Solar Budget. About 70% of all incoming radiation is absorbed, mostly by the land and sea, but also by the air and clouds. The ultraviolet wavelengths are absorbed by oxygen and ozone; while the longer infrared wavelengths are absorbed by carbon dioxide and water vapour. An important transformation takes place during this process. The wavelengths of the terrestrial radiation going out are longer than those coming in – too long to bounce back through the atmosphere and into space, which maintains the balance between heat gains and losses.

Because of the curve of the Earth, the radiant heat that falls on the 'fattest' part of the Earth, around the Equator and Tropics, is more intense than that which falls on the poles. These high levels of radiation in the Tropics heat the lower atmosphere which causes the air to rise and move to higher latitudes, where it then cools and sinks. This circulation which takes place in different cells between the Equator and poles, combines with the spinning of the Earth, produces global wind patterns. The same solar radiation also makes the seas and oceans warmer in the tropical regions: the warm water is then moved towards the poles by wind-driven currents. By transferring heat from the Tropics to the poles, oceanic and atmospheric circulation has a profound effect on the Earth's climate.

ABOVE: *El Niño is a reversal of the nomal flow of the Southern Equatorial Current, bringing dramatic weather changes.*

ABSORPTION VARIABLES

The amount of sunlight reflected from a surface, relative to the incident amount, is called the albedo. Bright surfaces have albedo near unity, and dark surfaces have albedo near zero. The DHR refers to the amount of spectral radiation reflected into all upward directions through an imaginary hemisphere situated above each surface point. The 'directional' part of the name describes how, in the absence of an intervening atmosphere, light from the Sun would illuminate the surface from a single direction (that is, there is no diffuse skylight, hence the name 'black-sky' albedo). To generate this product accurately, it is necessary to compensate for the effects of the atmosphere, and MISR's multi-angle retrieval techniques are used to screen clouds and account for the light scattered by airborne particulates (aerosols).

Different surfaces reflect or radiate back different amounts of radiation: think how pale colours reflect light and heat but dark colours absorb them. The percentage of the radiation reflected back by a surface is the albedo of the surface. A fresh covering of snow on the ground is highly reflective, which is why skiers wear sunglasses or goggles, and has an albedo of 75-95% (usually written as 0.75-0.95). Bright clouds in the sky are also very reflective, with an albedo of 0.70-0.90. Surfaces with a high albedo absorb little heat: dark surfaces absorb surface radiation much more. For example, a forest which has an albedo of 0.10-0.20 absorbs 80-90% of the radiation that falls on it. Cut the forest down and cover it in cement and the albedo rises to about 0.22, a change that can affect the local climate as increased radiation reduces convection in the air which, in turn, affects the amount of cloud cover.

BELOW: *Global models of the Earth system need accurate measurements of how much solar energy is reflected and absorbed by surfaces because this energy drives processes such as plant photosynthesis, snow melt, and longwave reradiation. These images from the Multi-angle Imaging SpectroRadiometer (MISR) provide global, seasonal summaries of a quantity called the Directional Hemispherical Reflectance (DHR), also sometimes referred to as the 'black-sky' albedo.*

The Structure of the Atmosphere

The atmosphere is among the Earth's most precious possessions. Without this protective blanket of gases and particles, the Earth could not sustain life. The Earth's atmosphere is relatively thin in comparison with the diameter of the Earth – if we could shrink the Earth down to the size of a ball 1 m (3 ft) in diameter, the layer of atmosphere would be just 2 mm (1/16 in) thick – yet it is thick enough to protect us from the deadly rays and objects that originate in outer space.

The atmosphere is composed of distinct layers which are quite uniform in their chemical composition, but their density (or mass per unit volume) decreases with altitude. As a result, the air pressure exerted on the Earth by the air also decreases with altitude. Most of the atmosphere's mass lies in the very lowest levels: about 99% lies within 30 km (18 miles) of the Earth's surface, but even at 80 km (50 miles) up, the atmosphere is still dense enough to intercept incoming solar rays and scatter them in all directions.

RIGHT: *The atmosphere sustains life on Earth.*

DRY AIR

Air makes up 98% of the weight of the atmosphere: water vapour and airborne particles make up the remaining 2%. Dry air is a mixture of three main gases with relatively constant volumes: nitrogen accounts for 78.09%, oxygen makes up 20.95% and argon 0.93%. In addition, there are minute traces of carbon dioxide, neon, helium, methane,

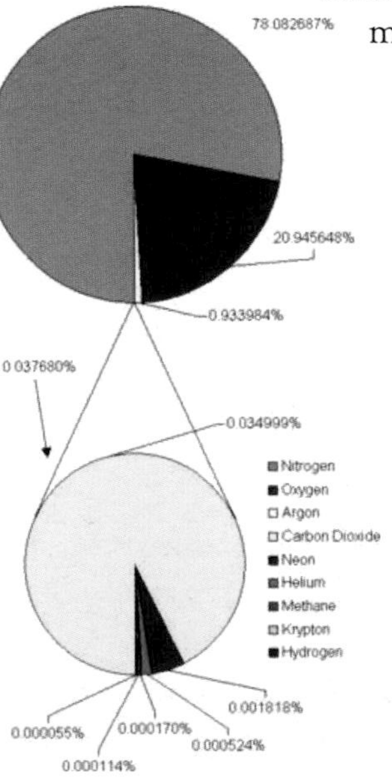

LEFT: *Nitrogen, oxygen and argon are the three main gases ofdry air. The chart shows traces of the other gases present in the atmosphere.*

LEFT: *Smog contains high concentrates of smoke and other pollutants.*

hydrogen, krypton, xenon, nitrous oxide, ozone and radon which make up 0.03%. The only gases that vary in volume to any significant degree are carbon dioxide and ozone: according to scientists, it is the fluctuations in the relative volumes of these two which could have far-reaching implications for the world's climate. Because of industrialisation, the air increasingly contains small quantities of gases which come from the surface of the Earth – ammonia, nitrogen dioxide, hydrogen sulphide and sulphur dioxide.

H^2O

Dry air is so-called because it omits a significant component: water. The water in the atmosphere exists in its three usual states: as water vapour, the colourless, odourless, gaseous form of water that is always in the air and which is the most plentiful; as liquid water, which is the chief component of clouds and occurs as droplets of various sizes; and in very small amounts, in its solid form, as tiny ice crystals which are often to be found mixed with water droplets. Most substances can exist in all these three 'states' but water is truly

BELOW: *Water is remarkable in that it exists as a vapour, a liquid and a solid form (as ice) at different temperatures on the Earth's surface.*

remarkable because it exists in all three states at temperatures that are experienced on the Earth's surface. In winter the surface of a pond may freeze over in places and become ice, but water is also present as a liquid in the pond and as a vapour in the air above it. With water, it is possible to have all three states present at the same time and in the same place.

ABOVE: *storm clouds*

The water in our atmosphere accounts for only 2% of global water stocks, which is estimated at 1.4 billion cubic kilometres (1/3 billion cubic miles) of water – that's all the water contained in the Earth's oceans, seas, ice sheets, glaciers, lakes and underground reservoirs. Nevertheless, this 2% performs the same sort of function as the glass roof of a greenhouse because it lets in incoming light waves and ultraviolet waves, but traps outgoing infrared rays. The equatorial regions are therefore a bit like giant furnaces, building up more heat than they radiate back into space. The heat produced in the equatorial regions ultimately ends up at the polar regions where it escapes into space – but not before it has stayed long enough in the northern and southern latitudes to keep the average temperature of the Earth at a constant 14°C, and made the zones between the Equator and the poles temperate.

Importantly, water in the atmosphere not only traps and bounces the infrared radiation back to Earth, it also stores energy. This energy is latent (or hidden) heat. Latent heat is heat absorbed or released by any substance that undergoes a change of phase without a change in temperature. In meteorology, latent heat applies to the three phases or states of water and to the relevant amounts of energy involved in the transition from one state to another – fusion (ice to liquid), sublimation (ice to water vapour) and vaporisation (liquid to water vapour). Water vapour moved by winds is one of the atmosphere's most important 'vehicles' for transporting heat. The latent energy in water vapour is converted back into heat when the vapour becomes a liquid: each time the air is cooled sufficiently, the water vapour in the air is condensed and forms clouds composed of tiny water droplets and the heat is released into the atmosphere. It is the release of this latent energy that fuels the formation of huge storm clouds.

AEROSOLS

Aerosols form the last element that makes up the atmosphere. They are microscopic-sized, solid particles of dust, volcanic ash, sand, pollen, grains, sea salt and smoke. Despite their small size, aerosols play a key role in the atmosphere. Their small sizes means that these solid particles remain temporarily suspended in the atmosphere: large particles fall to Earth in a few minutes, small particles such as smoke within a few hours. Some particles fall to the ground, some collide with surfaces as they are carried horizontally by the wind, while some particles – particularly dust, smoke and sulphates – get 'dissolved' when the water vapour in the atmosphere condenses on them, forming 'acid rain'. Other aerosols are beneficial and reflect sunlight back towards space, thereby partially offsetting the effects of global warming.

	2000
Dec - Feb	MISR obser began 24 Fe
Mar - May	
Jun - Aug	
Sep - Nov	

MISR aero

RIGHT: *Five years of atmospheric aerosol data are available as global aerosol maps from NASA's Multi-angle Imaging SpectroRadiometer (MISR). These 19 global panels show the seasonal-average distribution of atmospheric aerosol amount across Africa and the Atlantic Ocean. The measurements capture airborne particles in the entire atmospheric column, for sub-visible sizes ranging from tiny smoke particles to 'medium' dust (about 0.5 to 2.5 microns). Such particles are produced by forest fires, deserts, volcanoes, breaking ocean waves, and urban and industrial pollution sources.*

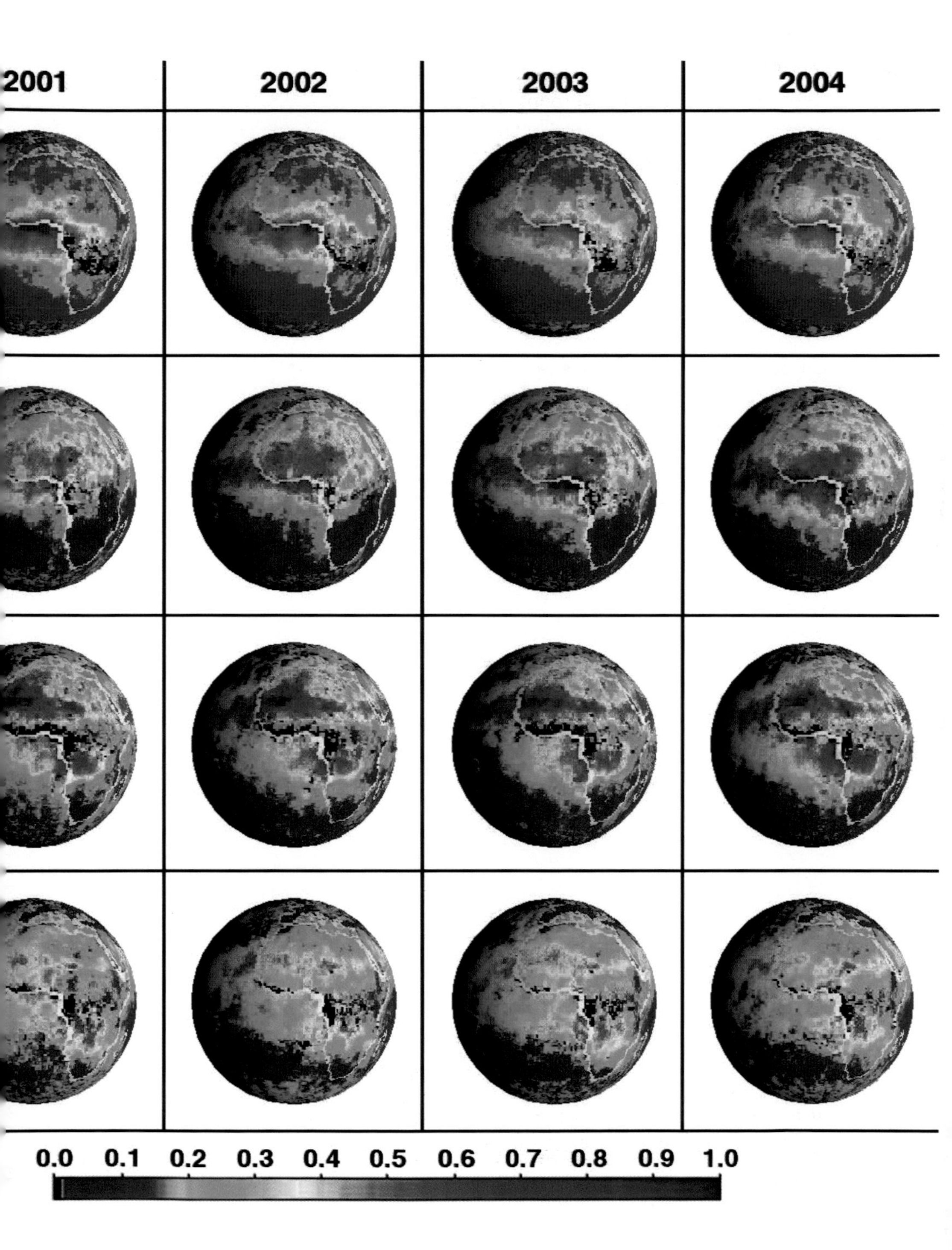
2001
2002
2003
2004
0.0
0.1
0.2
0.3
0.4
0.5
0.6
0.7
0.8
0.9
1.0

LAYERS IN THE ATMOSPHERE

The atmosphere consists of a stack of distinct layers. The first and lowest layer is the troposphere. At the poles, the troposphere extends to about 8 km (5 miles), while at the Equator it extends about 16 km (10 miles). This is the layer where clouds form, where air moves horizontally and vertically and is thoroughly mixed – in short, it's where the 'weather occurs'.

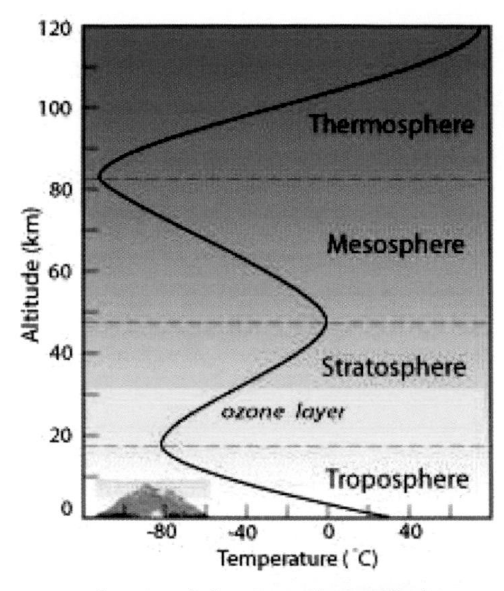

ABOVE: *Layers of the atmosphere. The lowest layer, the troposphere, is where life on Earth exists and where the weather occurs.*

The troposphere

In the troposphere, temperature decreases with altitude – the higher you go the colder it gets. On average the temperature drops at a rate of 6.5°C (43.7°F) for every 1,000 m (3,300 ft). In mid-latitude regions such as Britain and Europe, where the upper boundary of the troposphere, called the tropopause, varies in height from 9-12 km (5.5-7.5 miles), the temperature is around -55° to -60°C (-67° to -

76°F). The tropopause marks the upper level of the troposphere where air can cool no further. This is because the air above is no denser, so it rises no further.

Air pressure

Like temperature, air pressure also decreases with altitude: above 4,000 m (13,200 ft) there is not enough oxygen in the air to sustain vigorous activity, which is why mountain climbers often have to carry oxygen supplies with them. While the atmosphere around us may seem insubstantial, it does have a mass: held to Earth by gravity, the atmosphere weighs around 5,600 million tonnes (6,000 million tons)!

At sea level, the annual mean pressure is 1013.2 mbar (millibars) which is related to an air density of 1.23 kg/m3. In other words, a human standing at sea level has an atmospheric pressure of 10-20 tonnes (12-25 tons) being exerted on them! We aren't crushed by this enormous pressure because our own inner body pressure, pushing outwards, equalises the atmospheric pressure pushing in. Climb up to the top of the world's tallest building, the Petronas Towers in Kuala Lumpur, Malaysia, at 452 m (1,480 ft) above sea level, and the air pressure drops to 958.0 mbar and the air density is 1.17 kg/m3. At the top of the world's highest mountain, Mount Everest in the Himalayas, at 8848 m (29,000 ft) high, the air density is a mere 0.48 kg/m3 and the air pressure has dropped to 315.0 mbar. When you fly in a jumbo jet, its cruising height is around 11,000 m (36,000 ft) – right on the edge of the troposphere. Not only is the temperature outside a cool -56°C (-69°F), the air is extremely thin at only 0.36 kg/m3 and the air pressure is 225.0 mbar, which is why aircraft cabins are pressurised.

LEFT: *The summit of Mount Everest is well within the troposphere.*

The stratosphere

It was the French pioneer of weather balloons Leon Teisserenc de Bort who, in 1898, noticed that at around 10 km (6 miles), the temperature not only stopped falling, it actually rose slightly. Bort had discovered the stratosphere, the second layer of the atmosphere, which is about 30 km (18 miles) thick. It was in the virtually dust-free and almost cloudless stratosphere that the supersonic airliner Concorde flew (at around 18 km (10 miles) high, where the air density was a mere 0.12 kg/m3 and the air pressure just 70.0 mbar). In the stratosphere, the temperature increases with altitude: starting at about -56°C (-69°F) at around 20 km (12 miles), until it reaches 0°C (32°F) at the edge of the stratosphere (the stratopause) at a height of 50 km (30 miles) from the Earth' surface. This warming is due primarily to the presence of the ozone layer which lies some 25 km (16 miles) above the Earth's surface.

Nacreous clouds

Although the stratosphere is dry, on rare occasions clouds may form in it. These clouds are called nacreous or 'mother-of-pearl' clouds and are believed to be made of ice crystals which form smooth crests of vertical waves in the lower and middle stratosphere. Nacreous clouds occur up to around 30 km (18 miles) and are lenticular (lens-like) in form. Usually nacreous clouds are stationary and

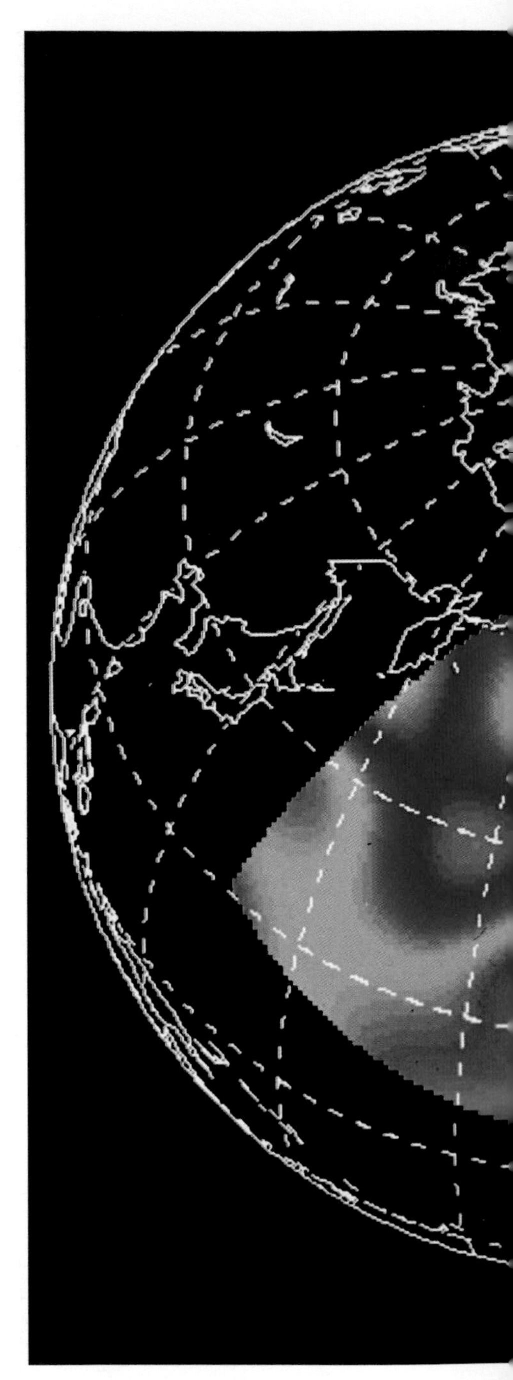

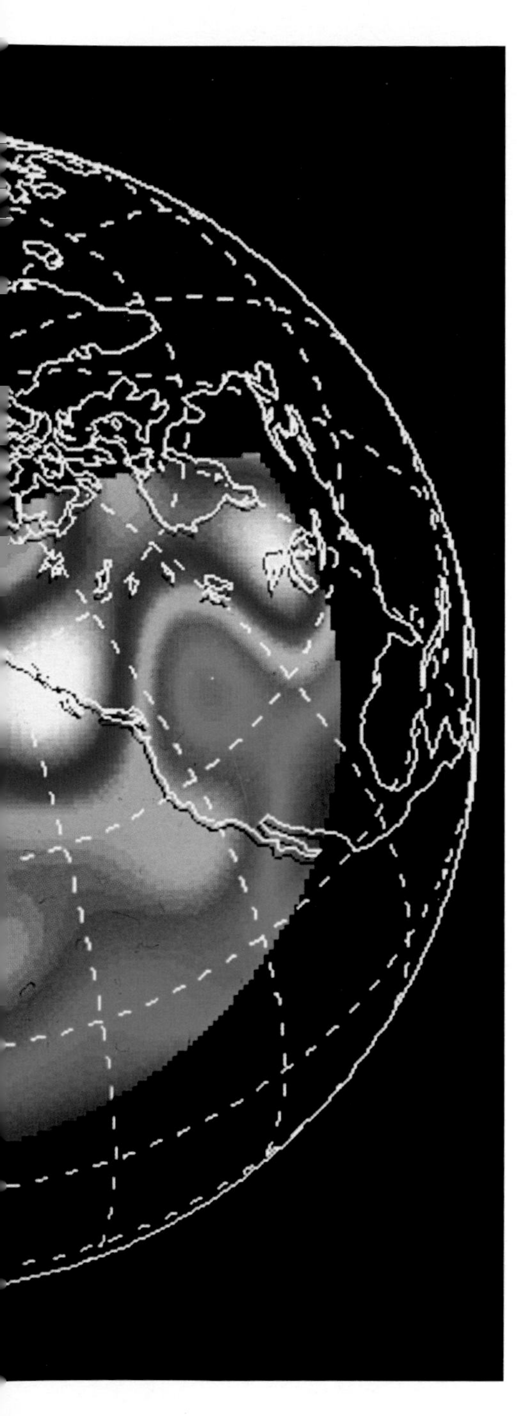

brilliantly iridescent – just like the mother-of-pearl that gives them their name. The colours, most often red and green but sometimes blue and green, are caused by very small cloud particles defracting the sunlight, rather like 'bending' a beam of light through a glass prism to separate out the colours of the spectrum in a 'mini rainbow'.

The ozone layer

Ozone is a toxic and highly reactive gas that is a rare form of oxygen – it has three atoms of oxygen (O_3) rather than the usual two (O_2) that we breathe. Ozone is, nevertheless, vital to life on Earth. It screens out the short-wave, ultraviolet solar radiation which carries high levels of energy – the solar radiation that causes sunburn and can cause permanent damage to living cells. The ozone molecules act as a 'screen', absorbing this radiation and re-emitting it as heat – hence the rise in temperature in the stratosphere. The ozone layer has been stable for thousands of years, until the 1920s, when scientists synthesised CFCs (chlorofluorocarbons) – non-flammable gases used in solvents, industrial cleaners, propellants in aerosol sprays and as coolants in

LEFT: *The ozone layer is a natural part of the atmosphere that screens out short-wave ultraviolet radiation.*

NEXT PAGE: *A familiar view of the top of middle level clouds, often seen from an aeroplane at 6,000 m (20,000 ft)*

refrigeration units and air conditioning. When CFCs escape into the atmosphere, they can destroy ozone.

In 1983, scientists with the British Antarctic Survey discovered an ozone 'hole' over the South Pole. In 1987, aware of the imminent dangers, several countries signed the Montreal Protocol, which committed industrialised nations to phasing out CFCs (this was followed by a further agreement in 1990 to include most developing nations to encourage the use of 'ozone-friendly' products). As a result, CFC levels in the atmosphere have reached a plateau and may even be falling, but how long it will take for the ozone layer to repair itself is another question. Although scientists have tentatively suggested it may not be until 2050 that the ozone layer is repaired, some CFC molecules only 'hang around' in the atmosphere for 45 years; others have a lifetime of 500 years.

Mesosphere and thermosphere

The next layer up in the atmosphere is called the mesosphere, about 50 km (30 miles) above the Earth's surface. In the lower mesosphere – from 50-56 km (30-35 miles) above the Earth – the temperature remains constant, but above 56 km (35 miles) the temperature falls with height, down to about -80°C (-112°F) at the boundary (mesopause) around 80 km (50 miles) high. This cooling promotes a vertical circulation of air which can sometimes lead to the formation of clouds over the polar regions in summertime. These clouds, known as noctilucent clouds, occur at elevations of around 80 km (50 miles), high enough for them to be illuminated by the Sun when the Earth's surface is shrouded in darkness.

The low temperatures in the mesosphere are in stark contrast to those found in the next layer, the appropriately named thermosphere. The thermosphere extends from an altitude of around 80 km (50 miles) up as high as 650 km (350 miles). In the lower part of this layer the temperature remains constant with height, but it increases rapidly above 88 km (55 miles) and in the upper reaches it reaches in excess of 700°C (1290°F) because of the absorption of ultraviolet radiation. In the thermosphere, the gases separate out according to their molecular weights. The upper layer of the thermosphere – the thermopause – gradually extends into the exosphere, or 'interplanetary space' where Space Shuttles and weather satellites orbit – at 1,000 km 620 miles) and 36,000 km (22,300 miles) respectively.

RIGHT: *The Space Age has allowed meteorologists to place geostationary and polar-orbiting satellites into space in order to transmit vital images of weather formations in the Earth's atmosphere.*

THE AIR IN MOTION

IN ANCIENT TIMES, the wind was feared and respected as a superhuman force that could mete out rewards and dire punishments. The Greeks distinguished between wind directions and associated them with different gods, one for each of the cardinal points: Boreas was the god of the north wind; Apeliotes, the east wind; Notos, the south wind; Zephryos, the west wind. The philosopher Aristotle added four intermediate directions, northwest, southeast and so on, and his eight-sided arrangement was the basis for the Tower of the Winds, an octagonal tower topped by a wind vane that was built in Athens, Greece in the first century BC.

Later, the Romans added 16 further points to the 'wind rose' bringing the total number of wind directions to 24. So inspiring were the winds that some were given their own names which are still with us today: the Mistral, the Sirocco, Chinook, Bora, Meltemi, and Leveche. People always guessed that the wind 'carried messages' about further weather: they could watch the way trees bent or the way smoke drifted; a wind out of a certain 'quarter' would mean fair weather, out of a different quarter, it would bring storms. But what the wind was and how it worked, would take scientists several hundred years to understand fully.

LEFT: *Boreas, Greek god of the north wind.*

GLOBAL ATMOSPHERIC CIRCULATION – THE EARTH'S AIR-CONDITIONING SYSTEM

The Earth is in constant motion and so, too, is the air around us. If the atmosphere was perfectly stable and inert, it would revolve in unison with the Earth. But in such a world there would be no winds apart from what are known as convection currents: air heated in the Tropics would rise and flow towards the polar regions where it would cool and become dense, so sinking and returning back to the Equator to replace the rising warm air. If this was the case, the only 'winds' we would have on Earth would be winds blowing from north to south from the poles to the Equator and consequently, in the so-called temperate latitudes, the weather would be perpetually cool, constantly damp, and always overcast.

On a global scale, the movement of air is the circulation of the atmosphere which transports warmth from equatorial regions to areas of high latitudes, and returns cooler air back to the Tropics – like a huge air-conditioning system! Two factors are largely responsible for 'moving things along' near to the surface of the Earth: the uneven spread of the Sun's energy over the globe, and the fact that the Earth is rotating.

Air acts a bit like a fluid when it is heated by the Sun's energy: the molecules absorb energy which allows them to move faster and so they move further apart and the fluid – or air – expands. Imagine a saucepan of water set on a stove to boil: as the water heats up, small bubbles start to appear. As the water boils, the bubbles get bigger and faster and can 'expand' to overflow the pan. Expansion makes the air less dense because a given volume contains fewer molecules than before. Denser air, like fluids, is heavier and sinks underneath the less dense air, pushing it upwards. So as warm air rises, air pressure decreases because there is a smaller amount of air above it to press it down. When the air reaches a level where its density is the same as the air immediately above it, it stops rising. If the air then subsides, the pressure on it increases. It is these changes that give rise to the different 'cells' of the Earth's circulation system.

Hadley cells

It was the English astronomer Edmund Halley (see page 29) who in 1686 first published a map showing the general circulation of the winds and noticed that changes in them were directly linked to changes in pressure – although he didn't quite understand why! It was not until 1735 that George Hadley factored in the Earth's rotation which accounted for the trade winds: Hadley showed that warm air actually sank back towards the Earth's surface before it reached the poles. Hadley realised that there was more than one circulation loop or 'cell' at work: there were tropical Hadley cells (where the warm air rises at the Equator, flows towards the poles, cooling down on the way) and polar Hadley cells (where cold, heavy air flows away toward warmer areas and as this air comes in contact with the ocean surface, it gradually heats up, then rises, and changes direction).

BELOW: *The movement of warm air.*

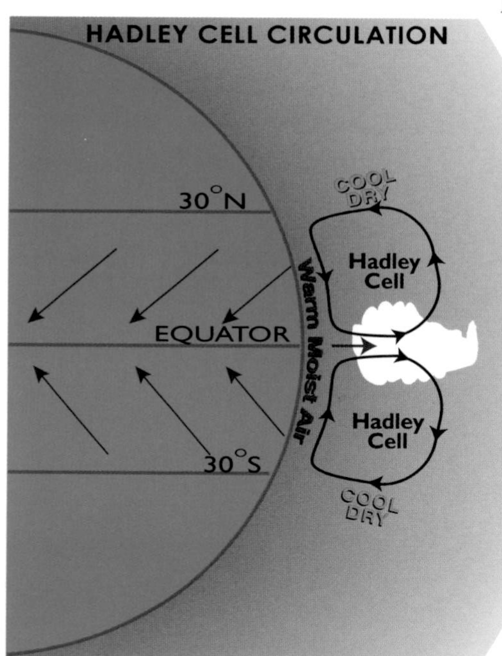

The Coriolis effect

A major discovery regarding the circulation of the atmosphere came in 1835 and was made by the French mathematician and mechanical engineer, Gustave-Gaspard de Coriolis. Coriolis discovered that air flowing towards or away from the Equator invariably follows a curved path that swings to the right in the northern hemisphere, and to the left in the southern hemisphere. Known initially as the Coriolis Force, meteorologists abbreviate it as CorF, but it is known today as the Coriolis effect because no 'force' is involved. The Coriolis effect occurs because the Earth is rotating anticlockwise about its axis, so, as the air moves across the Earth's surface, the surface itself is also moving – but at a different speed. All points on the Earth's surface complete one revolution in 24 hours: a point on the Equator (the 'fattest' bit of the Earth) travels further and faster than points at higher latitudes. This rotation causes the air to shift to the right (in the northern hemisphere) or left (in the southern hemisphere). So, the magnitude of the Coriolis effect depends on latitude and the speed of the moving air, which makes the CorF greatest at the North and South poles.

You can test the Coriolis effect for yourself using a record player turntable, a cardboard disc with a map of the northern hemisphere on it, a pencil or pen, and a ruler. Place the 'disc' with the North Pole over the spindle and spin the turntable anticlockwise (the direction of the Earth's rotation in the northern hemisphere). Put the ruler on top of the spindle and draw a line along it as the disc spins. Although you will have drawn a straight line along the edge of the stationary ruler, the line described will be an arc when the disc stops turning. You can carry on drawing lines from the Pole (spindle) to the Equator (at the rim of the disc) and they will all curve to the right – or the west!

Ferrel cells

In 1856, an American meteorologist William Ferrel put forward a circulation pattern for each hemisphere that was based on three cells: Hadley's polar and tropical cells and his own middle latitude cell. The Ferrel cell is a region of rising air at around 50°-60°N and 50°-60°S of the Equator – situated between the tropical and polar Hadley cells. Ferrel recognised that in the Hadley tropical cell, warm air rose and moved polewards at high altitudes and was deflected by the Coriolis effect as a westerly flow. Above 30° north or south of the Equator, the air cooled and subsided, with some returning to low latitudes as the easterly flow of the trade winds, with other air moving polewards as westerlies. The polar Hadley cell was where the cold dense

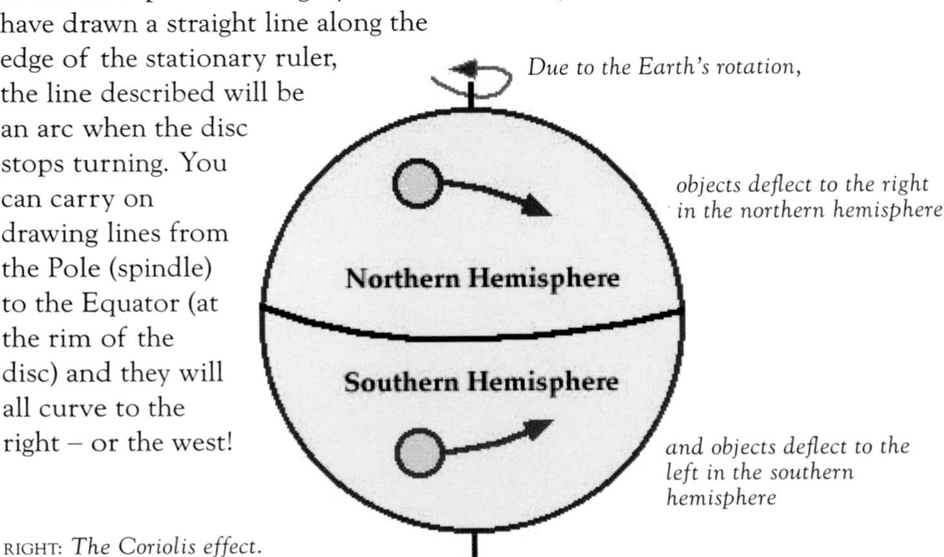

RIGHT: *The Coriolis effect.*

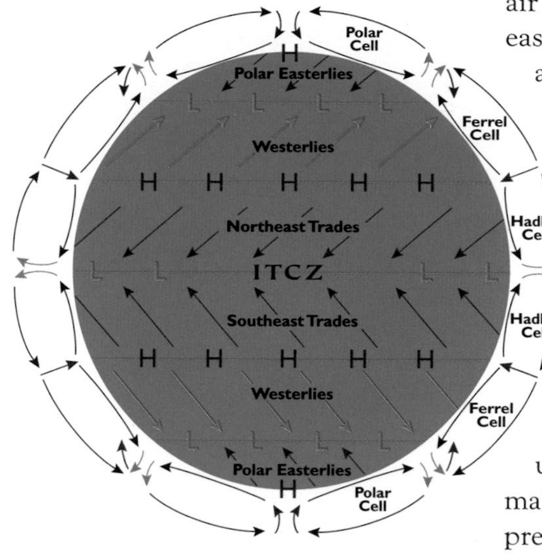

ABOVE: *The circulation of the atmosphere according to William Ferrel in 1856.*

air returns to lower latitudes as easterly winds. Between the polar and tropical Hadley cells are the additional cells proposed by Ferrel, where the outflowing surface winds clash and circulate. Air rises to the troposphere and then divides, with some air flowing to the poles, and some towards the Equator in the form of westerly winds.

Ferrel's model was subsequently used to indicate where rising air makes for significant amounts of precipitation, and where sinking air creates high-pressure zones with arid conditions, notably the hot deserts.

Ferrel's theoretical investigations of atmospheric circulation, which culminated in his proposed mid-latitude cells, predates Christopher Hendrik Buys-Ballot's empirical discovery in 1857. Buys Ballot's law states that if, in the northern hemisphere, a person faces downwind – that is, you turn your back to the wind – then low pressure is on the left and high pressure is on the right. In the southern hemisphere, it's the opposite way round.

Prevailing winds

Although it seems that the wind may blow from any direction, when the directions are compared at a particular place over a period of time, it usually turns out that the wind blows more often from one direction than from others. These are known as prevailing winds and the general circulation system of the atmosphere produces prevailing winds all around the world. In the Tropics the winds that are part of the Hadley cells that blow towards the Equator and are deflected to the right are the northeasterly and southeasterly trade winds. Air flowing away from the poles and deflected to the right forms the polar easterlies. And in-between the two are the winds of the Ferrel cells producing westerly winds. All over the world, the force of the easterly winds is balanced by that of the westerlies: if this wasn't so the Earth would either speed up or slow down!

In sandy desert and polar regions, it is the prevailing winds that produce the sand dunes and snow drifts with their characteristic crest-shaped tops. Grains of sand or snow are blown up a slope and then tumble down the far side. Knowing about the prevailing winds was vital in the past to navigators who depended on them to cross the oceans without getting becalmed in the Doldrums. Today, the prevailing winds partly dictate the location of structures such as airport runways.

ABOVE: *It is the Coriolis effect that makes the winds spiral in tropical cyclones.*

The trade winds and the Intertropical Convergence Zone

The trade winds that blow over the great ocean areas of the Tropics and

RIGHT: *Prevailing winds form the characteristic crests on the tops of sand dunes. Grains of sand are blown up the slope and tumble down the other side.*

subtropics are large-scale convection winds – that is, they are 'powered' by an imbalance in temperatures, in this case the difference between the temperatures at the two poles and the Equator. The trade winds are known for their strength and constancy: they blow strongly all the time throughout the year, and always in the same direction. In the northern hemisphere they are northeasterlies, in the southern hemisphere, the trades are southeasterlies. Without the trade winds, European explorers would never have discovered the 'New World'. European traders also made use of the trade winds for the 'triangular trade' between Europe, Africa and the Americas, sending their ships laden with goods on northeasterly trade winds to Africa, where they were exchanged for slaves. Loaded with their human cargoes, the ships would then ride the trade winds to the Americas, barter the slaves for cotton or sugar, then return on the prevailing westerly winds to Europe to begin the cycle again.

It's easiest to think about the trade winds as giant 'conveyor belts' only a bit more sophisticated! As they reach each other, the trade winds slow down dramatically and enter what is called the Intertropical Convergence Zone (ITCZ). Sailors know this area as the Doldrums: it lies within approximately 5°-10° either side of the Equator and is

where the winds are very light and variable, causing sailing ships to be becalmed. The winds are light because there is very little horizontal movement of air: the Sun's blazing heat lifts it almost straight up! This hot air rises up and branches out like a fountain, some north, some south, forming upper levels of air currents known as the anti-trade winds – literally stopping ships from trading!

When these anti-trade winds reach about 25° latitude north or south, they divide again. One stream continues towards the poles and forms the upper level winds known as westerlies. The other begins to descend back to Earth where it piles up in the region of the 30th parallel, creating a fair-weather zone of calm air which sailors call the Horse Latitudes –presumably on account of all the horses that died and had to be thrown overboard from becalmed Spanish ships on their way to the West Indies. When the falling air of the Horse Latitudes reaches the Earth's surface, it too divides: one stream returns to the Equator, the other heads for the poles. It is the equatorial current that forms the trade winds. These, in time, replace the air rising over the Doldrums, thus completing the equatorial convection system – the Hadley equatorial cell.

LEFT: *Trade winds carried explorers to the New World.*

Westerlies

LEFT: *Clouds caught in the high jet stream*

Thanks to the Coriolis effect – the spin of the Earth from west to east – the trade winds are steered and made to veer off a headlong north-south course. The air currents that move towards the poles from the Horse Latitudes are deflected eastwards and merge into the prevailing westerly wind – the second great wind system of the global circulation system.

The westerlies circle the Earth in waving bands between the 30th and 60th parallels. Even at ground level, westerly winds can be the wildest of all the persistent winds: they whip up storms in the North Atlantic and the English Channel while in the southern hemisphere, because there are no land masses to interrupt their flow, they blow at gale force across thousands of kilometres of open ocean, driving before them storm clouds and rolling the sea into huge waves up to 18 m (60 ft) high. These westerly winds are most often found in the latitudes between 40° and 50° south – the famous 'Roaring Forties'.

Jet streams

Embedded in the main currents of westerly winds, roughly 10 km (6 miles) up, are the jet streams. Like all the winds, jet streams are invisible, but sometimes their effects can be seen when their passage creates a disturbance in high-altitude clouds. Jet streams were first identified by

MAIN IMAGE: *Aircraft use the east-bound jet stream to speed the passage from Europe to America.*

the Swedish-American meteorologist Carl-Gustav Rossby in the 1920s, but were first experienced by American fighter pilots in World War II. They are narrow, winding bands of strong wind in the upper troposphere blowing at around 400-600 km/h (200-300 mph), and they are the winds which help to 'push' east-bound airplanes across continental America two hours faster than they can make the same flight in the opposite direction! The highest speed recorded was 656 km/h (407 mph), above South Uist in the Outer Hebrides, Scotland, in December 1967. Jet streams mark the boundary between air masses of different temperatures – they are also known as thermal winds – and are closely associated with weather fronts and depressions.

Polar jet streams flow in temperate latitudes, between 30° and 50° north and south all year round. The subtropical jet stream is at about 30° north and south all year round. The jets streams blow with cold air on their left in the northern hemisphere and cold air on their right in the southern hemisphere: both therefore blow from west to east. In summer, however, there is a jet stream that blows from east to west at about 20° north of the Equator, crossing Asia, southern Arabia, and northeastern Africa.

Polar easterlies

As the westerly winds near the poles, they lose their speed and force: at the 60° latitudes, they brush up against the bands of winds that make up the third winds of the global circulation system, the polar easterlies.

The polar easterlies begin as masses of cold air accumulated in

areas of intense cold, close to the North and South poles: in the northern hemisphere these cold spots occur in Siberia, around the towns of Vekhoynask and Oymyakon where winter temperatures fall to -68°C (-90.5°F), and in northwestern Canada at the town of Snag, where winter temperatures are around -63°C (-81.5°F). Greenland is another 'cold spot' in the northern hemisphere, while in the south, in Antarctica, the Russian weather station at Vostok has registered temperatures of -88.3°C (-27°F). These 'poles of cold' produce huge mantles of heavy, cold air that fan out and move towards the Equator but are deflected by the Coriolis effect to become polar easterly winds.

WHAT MAKES THE WINDS BLOW?

The interior clouds are thicker and likely to be more convectively active than the other clouds, causing much of the air near the centres of the arcs to rise. This air spreads out horizontally in all directions as it rises and continues to spread out as it begins to sink back to the surface. This pushes any existing small cumulus clouds away from the central region of convection.

As the air sinks, it also warms, preventing other small clouds from forming, so that the regions just inside the arcs are kept clear. At the arcs, the horizontal flow of sinking air is now quite weak and on meeting the undisturbed air it can rise again slightly. Although examples of the continuity of air, in which every rising air motion must be compensated by a sinking motion elsewhere, are very common, the degree of organisation exhibited here is relatively rare, as the wind field at

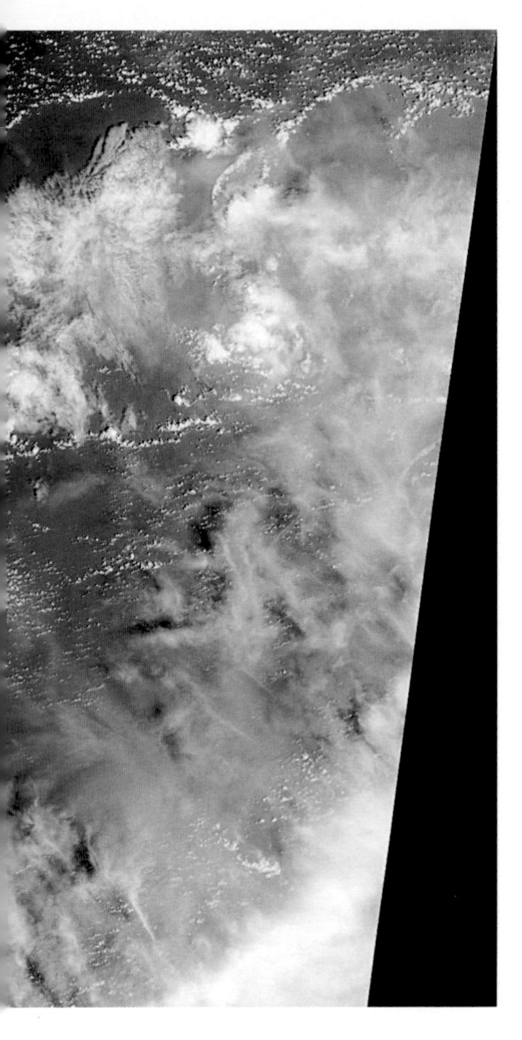

LEFT: *Small cumulus clouds in this natural-colour view from the Multi-angle Imaging SpectroRadiometer have formed a distinctive series of quasi-circular arcs. Clues regarding the formation of these arcs can be found by noting that larger clouds exist in the interior of each arc. The image was acquired by MISR's nadir camera on March 11, 2002, and is centred west of the Marshall Islands.*

different altitudes usually disrupts such patterns. The degree of self organisation of this cloud image, whereby three or four such circular events form a quasi-periodic pattern, probably also requires a relatively uncommon combination of wind, temperature and humidity conditions for it to occur.

The trade winds, the westerlies and the polar easterlies are the three major components of the global circulation system – the global air convection or 'air conditioning'. Constantly moving, transporting warmth from the Tropics to higher latitudes and returning cool air back to the Equator, the aim is to restore balance by redistributing pressure.

It was the French physicist Blaise Pascal (1623-1662) who demonstrated that the wind was just a mass of moving air. It's a little like when the tyre of a car bursts: the air that was held under pressure (about 13.5kg/cm^2 (30 lb./in^2) in the tyre rushes out into the surrounding air (which is on average around 6.68kg/cm^2 (14.7 lb./in^2). Exactly the same thing happens in the atmosphere: air looks for a way to escape from areas of high pressure to move to areas of low pressure, and the shift of the air creates wind. The greater the difference of the pressure between the two areas, the stronger the wind. Consequently, wind speed, and wind direction are dictated by the position of areas of high and low pressure.

Pressure and Wind Speeds

THE DIFFERENCE BETWEEN high and low pressure across the Earth's surface is the basic driving force that makes the wind blow. The speed of the wind is proportional to the pressure gradient – the rate of pressure change over a horizontal distance. This means that warm, rising air produces an area of low surface pressure, or cyclone. Dense, heavy, subsiding cool air forms an area of high surface pressure, called an anticyclone. The difference in pressure between the centre of a pressure system and the air outside it is the pressure gradient, and air flows along this gradient from areas of high pressure to areas of low pressure: the steeper the gradient, the stronger the wind. It's a bit like a skater making spins on ice – drawing the arms into the body means they turn faster. The closer the air get as it spirals into the low-pressure centre, the smaller the radius of the spiral. In a deep depression, drawing air from a large area towards a small centre makes larger winds, which is why cyclones and tornadoes – which have small 'cores'– generate such strong winds.

RIGHT: *Characteristics of a strengthening Category 3 Hurricane Lili are apparent in these images from the Multi-angle Imaging SpectroRadiometer (MISR), including a well-developed clearing at the hurricane eye. When these views were acquired on October 2, 2002, Lili was approaching the Gulf coast of the United States and rapidly strengthening toward Category 4 status. The storm's power reached its peak less than twelve hours later, and although it weakened overnight, this was still a dangerous system as it blew across the Louisiana coast on the morning of October 3. Lili was the first hurricane to make landfall in the United States since Hurricane Irene in 1999. Twenty-eight parishes in Louisiana were declared disaster areas, yet this hurricane fortunately caused much less damage than what could have resulted from an event of this magnitude.*

When the barometer is falling, the pressure gradient – the size of the pressure difference and the distance between areas of high and low pressure – increases. When the flow of air from high to low pressure accelerates and reaches speeds of 100 km/h (63 mph) or higher, it is officially designated a storm. By international agreement, the national weather services must issue storm warnings when wind speeds reach 10-11 on the Beaufort Scale because

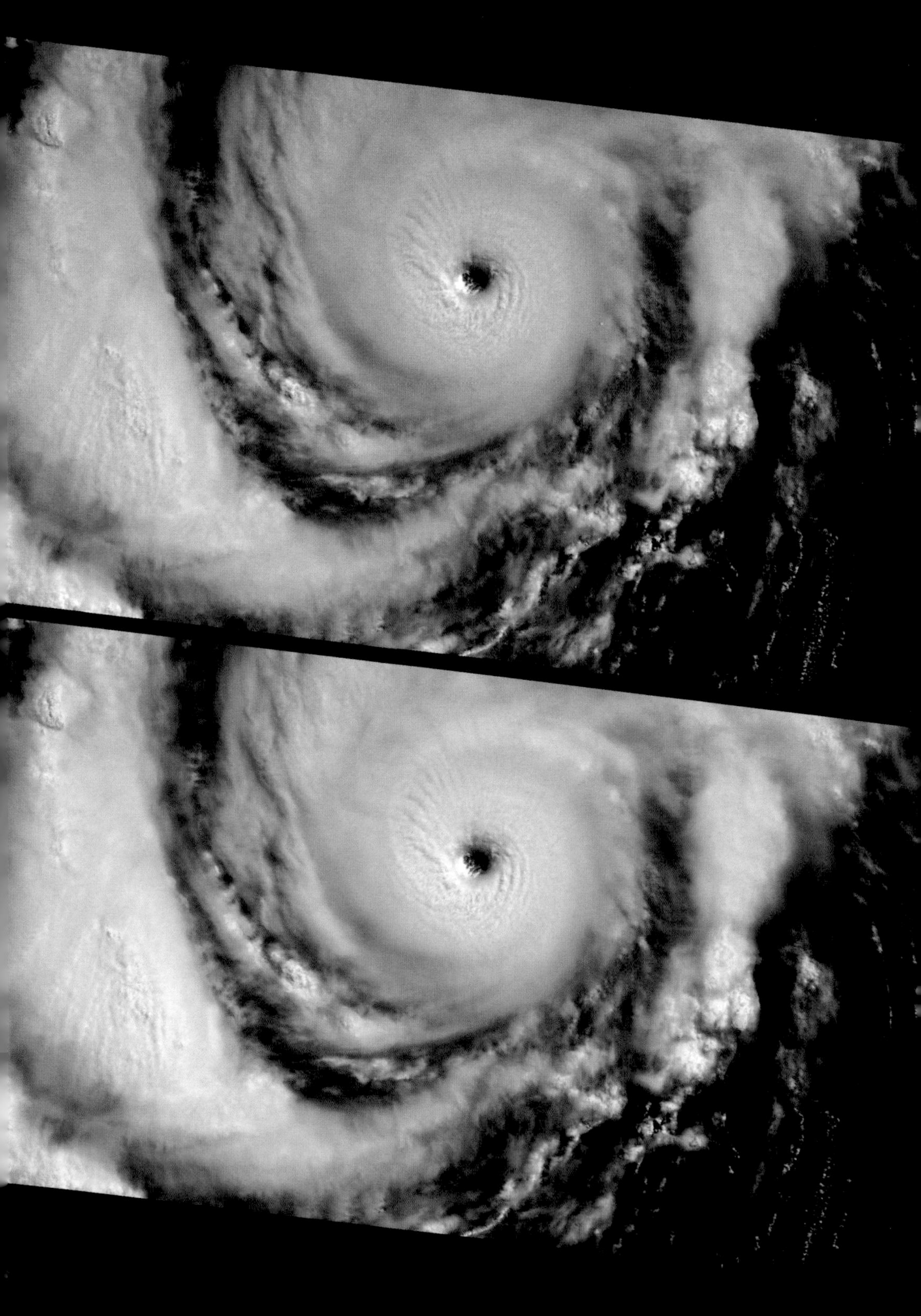

rough seas swept along by very high winds pose a threat to all shipping in the area affected.

THE BEAUFORT WIND FORCE SCALE

Devised by an English naval officer, Sir Francis Beaufort, in 1805, the Beaufort Scale was a scale of numbers originally intended for sailors who used visible features of the sea to judge wind strength. It was later modified for use on land by substituting features such as umbrellas, chimney pots and trees.

OCEANIC WINDS

The wind is stronger over the sea than over the land because there is less friction over water. Hills, buildings and other obstructions slow the wind over land. The distance that the wind blows without interruption is called a fetch. It is the long fetch over an ocean that affects and alters the air masses crossing it from a continent: the longest fetch is in the Southern Oceans in the Roaring Forties where, uninterrupted by land masses, the wind blows all

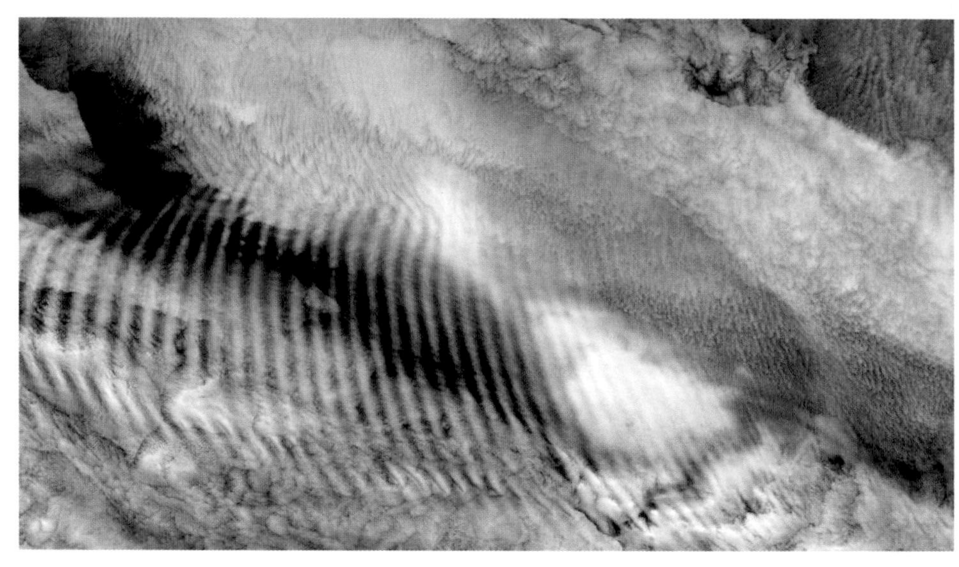

ABOVE: *In this natural-color image from the Multi-angle Imaging SpectroRadiometer (MISR), a fingerprint-like gravity wave feature occurs over a deck of marine stratocumulus clouds. Similar to the ripples that occur when a pebble is thrown into a still pond, such "gravity waves" sometimes appear when the relatively stable and stratified air masses associated with stratocumulus cloud layers are disturbed by a vertical trigger from the underlying terrain, or by a thunderstorm updraft or some other vertical wind shear. The stratocumulus cellular clouds that underlie the wave feature are associated with sinking air that is strongly cooled at the level of the cloud-tops -- such clouds are common over mid-latitude oceans when the air is unperturbed by cyclonic or frontal activity. This image is centered over the Indian Ocean (at about 38.9° South, 80.6° East), and was acquired on October 29, 2003.*

THE BEAUFORT SCALE

FORCE	DESCRIPTION*	EVENTS ON LAND	KM/H	MPH
0	Calm	Smoke rises vertically	under 1	0–1
1	Light Air	Direction of wind shown by smoke, but not wind vane	1–5	1–3
2	Light Breeze	Wind felt on face, leaves rustle, wind vane turns to wind	6–11	4–7
3	Gentle Breeze	Leaves and small twigs in motion, small flags spread	12–19	8–12
4	Moderate Breeze	Wind raises dust and loose paper, small branches move	20–29	13–18
5	Fresh Breeze	Small leafy trees sway, wavelets with crests on inland waters	30–39	19–24
6	Strong Breeze	Large branches move, whistling in telephone wires, hard to use an umbrella	40–45	25–31
7	Near Gale	Whole trees in motion, difficult to walk against wind	51–61	32–38
8	Gale	Twigs break from trees, difficult to walk	62–74	39–46
9	Strong Gale	Slightly structural damage to buildings, chimney pots, roof tiles and TV aerials removed	75–87	47–54
10	Storm	Trees uprooted, considerable damage to buildings, cars overturn	88–101	55–63
11	Violent Storm	Widespread damage to all types of buildings	102–119	64–74
12	Hurricane	Widespread destruction; only specially constructed buildings survive	120+	75+

** Official World Meteorological Organisation terms, 1964*

the way around the world. Continental air is dry and water will evaporate into it as it crosses the ocean. Evaporation would be very slow if the air was still, and the lowest layer of the air would become saturated. But, because the wind is constantly moving, it replaces the wet surface layer with dryer air from above and greatly increases the rate by which air gathers moisture, changing it from a continental air mass to a maritime air mass.

LOCAL WINDS

Differences in pressure are usually associated with large-scale weather systems, but they can also occur and produce winds on a smaller scale. Local wind systems are created by adjacent areas being unevenly heated by the Sun. During the day, as a hill or mountainside absorbs heat from the Sun, the air directly in contact with the surface begins to rise as it becomes warmer. As it becomes warmer, the air becomes less dense and lighter than the air in the valley below. The cooler air in the valley rushes in to take the place of the warm air and thereby creates a breeze, known as an anabatic (or valley) wind blowing up the slope. Hang-gliders and glider pilots soar by hitching rides on these rising thermal currents, and know that they cannot extend their time in the air beyond a certain hour: as night falls, the situation is reversed. With no sunshine to warm them, the cooling mountain slopes cause increasingly heavy, dense and cool air directly above it to slide down into the valley

BELOW: *Hang-gliders use rising warm air currents or 'thermals' to keep them aloft.*

and produce a breeze – called a katabatic (or mountain) wind – in the opposite direction, down the mountainside.

ABOVE: *Coastal regions have their own wind systems of onshore and offshore breezes.*

DAILY AND SEASONAL WINDS

In addition to anabatic and katabatic winds, there are daily winds such as land-sea breezes which occur at certain times of the day. Solar heating affects land and bodies of water differently: air over the land warms up and cools down faster than it does over the sea. These temperature contrasts generate land, or offshore, breezes and sea, or onshore, breezes. The coastal regions of oceans, and large lakes such as the great lakes between the USA and Canada, have their own local wind systems, and the onshore and offshore breezes provide the winds to power windsurfers and sailors across the water. Many of the daytime sea breezes, and their night-time counterparts, appear with clocklike regularity, especially along the coastlines of the Tropics and subtropics. They are so regular that many have local names: the virazon of Chile, the datoo of Gibraltar, the imbat of Morocco, the ponente of Italy, the kapalilua of Hawaii, and the various 'doctors' found in English-speaking tropical and subtropical regions, such as the Cape Doctor in Cape Town, South Africa

– so called because it brings 'relief' from the searing heat!

Further north, sea breezes tend to be seasonal, appearing in late spring and early summer. They spring up around 10-11 a.m. and subside around 2 p.m., dying out completely around 7 or 8 p.m.

Seasonal winds

Some winds are associated with particular seasons. The haboob, for example, is a dust storm that occurs in northern Sudan, usually late on summer days (haboobs are strongest in April and May, but occur every month except November). The wind direction may be north in winter, or east, southeast or south in summer. When a squall with strong down draughts moves over ground covered with loose sand and dust, it can raise a pillar of sand and dust to a great height – up to 1,000 m (3,300 ft) high. When several squalls merge, they can create a wall of dust up to 24 km (15 m) long which can advance at speeds of about 56 km/h (35 mph). In March 1998, a haboob passed over Egypt, parts of Lebanon and Jordan, reducing visibility to 180 m (600 ft) and causing the closure of Cairo Airport and the Suez Canal.

The sirocco

The oppressively hot sirocco most often occurs in the spring and blows dry, dust-filled air from the deserts of North Africa, northwards over the Mediterranean and as far north as the southern shores of Europe. As it crosses the sea, the sirocco gathers moisture so by the time it arrives on continental Europe, it's warm and wet and ready to dump its accumulated dust in the form of sandy, red rain – called 'blood rain' in Britain. Scientists estimate that each year 3 billion tonnes (3.3 billion tons) of dust is carried away each year from the Sahara and the Sahel by the wind. Some is carried so high into the atmosphere that it is circulated around the world and is deposited as far afield as Greenland and the Caribbean. Such deposits may have long-term implications: for example, the white ice of Greenland reflected sunlight and stayed frozen, but the dark dust which now settles on top absorbs the Sun's energy causing the ice to melt- and accelerating the raising of sea levels.

The Meltemi or Etesian winds

Between May and September over the Aegean Sea and the eastern Mediterranean blow the Meltemi or Etesian winds. They are generally strong northerly winds. Fiona Campbell, a marine meteorologist employed by the British Royal Yachting Association, spent four years measuring the weather in Athens for the British 2004 Olympic sailing team to see exactly how the wind behaved. Campbell predicted

ABOVE AND RIGHT: *The Santa Ana winds that typically blow through Southern California during late autumn and winter swept large amounts of dust and ash across the skies of San Diego and over the Pacific Ocean on November 27, 2003. The intense brush fires that had swept through the foothills of this region in October left soils exposed and vulnerable to such strong winds.*

that the Meltemi would be a mere sea breeze and the local topography would cause it to shift to the left (in Britain sea-breezes shift to the right). This was vital information for the British sailors: drawing on Campbell's conclusions, the British sailing team won two gold, one silver and two bronze medals.

Meanwhile, the harmattan blows from the north or northeast over West Africa to the south of the Sahara at any time of the year – but only during the day. The dry, dusty, but relatively cool, harmattan is part of the trade wind system, and in the wet summer season it tends to give way to the monsoon winds from the Gulf of Guinea and the Atlantic Ocean.

Monsoons

The most dramatic regional, persistent wind is the monsoon – a word derived from the Arabic *mausim* which means 'season'. It was first used to describe the seasonal winds of the Arabian Sea that blow

BELOW LEFT: *The southwest monsoon wind is associated with the rainy season: the arrival of these rains – often accompanied by heavy flooding – is vital to many countries' economies.*

for six months from the northeast, then reverse and blow just as steadily for the other six months from the southwest. The monsoon has always played a vital role in the economy of the Middle and Far East: it was monsoon winds that blew trading vessels from the east coast of Africa across the Indian Ocean to the rich Malabar Coast of India. In the first century AD, Arab sailors travelled northeast across the Gulf of Aden to the mouth of the River Indus. Three centuries later, monsoon winds took them all the way to China.

Monsoon-like winds occur in many parts of the world, but the best-known are the two Asiatic systems divided by the Himalayas: the east Asia monsoon, which is the predominant wind of mainland China and Japan, and the great South Asian monsoon, powered by the heating and cooling of the Indian peninsula jutting out into the Indian Ocean. The economies of many countries are at the mercy of the monsoon: for example, India's rice crop depends on the moisture brought from the Indian Ocean by the monsoon.

Orographic winds

All over the world there are persistent winds that derive their unique character from the local topography – the so called orographic winds. When moving air meets mountains or other geographical 'barriers' it can do one of two things: it can flow round them and through the gaps between them, or go up and over them. Funnelled into narrow valleys, airflow speeds up and causes the wind to blow from a particular direction. One such wind is the bora, a type that is also known as a fall wind or drainage wind. It is a massive 'river' of air that falls or drains from high, cold plateau regions on to a warm plain or coastal area. Under bright sunny skies, bora winds blow bitterly cold and furiously across the land. The most celebrated bora – the eponymous Bora – occurs in winter when there is low pressure over the Mediterranean and high pressure over central Europe. The Bora howls out of the Balkans and can at times, paralyse the Adriatic and Dalmatian coasts, bringing shipping to a halt.

Another bora-like wind is the French mistral which drains off the plateau of central France. The Greek geographer Strabo, writing in the first century AD, called the mistral 'an impetuous and terrible wind' that was capable of hurling rocks, stripping the clothes off peoples' backs and throwing men from their chariots. It is indeed a cold, dry and relentless wind that blows south from Burgundy in the spring and autumn, and is funnelled down the narrow corridor of the Rhone Valley. Squeezing all this air through a narrow funnel accelerates the airflow sharply resulting in winds roaring through the Rhone Valley at speeds of 75-100 km/h (47-63 mph), before it sweeps across Provence.

The fohn effect

Another kind or orographic wind is the trans-mountain wind such as the chinook, the fohn and the zonda. Rather than go around mountains and through valleys, the second option for moving air when it meets a mountain is to go up and over it. When air is lifted up by the sloping terrain, the increasing altitude causes it to cool and expand. The water vapour it carries reaches its dew point, condenses and forms clouds. This process can lead to some unusual phenomena: a single mountain such as Mount Washington, in New Hampshire, USA, is so tall that it wears a cloud on its summit more or less all the time.

At other times, conditions may be right for what is known as the fohn effect (named after the hot, dry winds that roar through the Alpine valleys), a variable phenomena which is responsible for small-scale weather patterns known as micro-climates. A fohn wind carries moisture-laden air high up mountainsides. As it rises,

RIGHT: *Physical barriers, such as mountain ranges, cause winds to act in certain ways: either the wind goes round the mountains and through gaps where it is funelled into narrow valleys, or it goes up and over the top.*

its temperature falls about 0.5°C (33°F) for every 100 m (330 ft) in elevation. Cooling causes the water vapour to condense, creating clouds which sometimes drop rain or snow on the windward side of the mountains. In the process, much of the water content in the air is effectively 'wrung out', so, once it has reached the top, the cooled dry air starts to rush down the other side, the leeward side, where it is warmed up by compression and begins to generate heat at the rate of roughly 1°C (33.8°F) for every 100 m (330 ft) it falls. The drier the air, the faster the rise in temperature. Consequently opposite sides of the same mountain can have radically different weather – at the same time! The most dramatic examples of the fohn effect are the Andean zonda, and the chinook, which blows down the eastern slopes of the Rocky Mountains. The fohn effect is not without its dangers: in winter, the rapidly warming air can cause snow to melt quickly and trigger avalanches.

BELOW: *Moisture-laden air rises as it is forced up the mountain; the air temperature drops and water vapour condenses into clouds.*

BIG WINDS

All of these winds are mere breezes, however, when compared to the really big winds – the tropical cyclones which are also known as hurricanes or typhoons, and the fiercest winds on Earth, tornadoes. Associated with extreme weather, hurricanes and tornadoes are 'created' by a number of contributing factors including air masses and weather systems.

Air Masses and Weather Systems

An air mass is a body of air that covers most of a continent or ocean and has uniform characteristics that it acquires from the Earth's surface below. Air over a continental land mass will be drier than air over the ocean, and it will be warmer in the summer and cooler in the winter because of the different rates at which the land and the sea warm up and cool down. Once it has taken on these characteristics, the body of air becomes an air mass.

The existence of air masses was discovered by the Norwegian meteorologist Vilhelm Bjerknes. In 1917 Bjerknes set up a series of weather stations in Norway and, using the data they provided, he was able to identify air masses separated by fronts, from which he was able to propose a theory describing the formation and dissolution of weather fronts.

Air masses extend from the surface of the Earth all the way up to the troposphere and have various names which describe their origins. Air masses that form over land are called continental, while those that form over the sea are maritime. Depending on the latitude in which they form, air masses may be arctic, polar, tropical or equatorial. These names can be further combined to give us six types of air mass: continental arctic, continental polar and continental tropical; maritime arctic, maritime tropical and maritime equatorial. Arctic and polar air are much the same at surface level, but they differ in the levels of the upper atmosphere. There isn't a continental equatorial type because there isn't a continental land mass in the equatorial region that is big enough to produce continental equatorial air.

LEFT: *Air masses are separated by fronts: when fronts meet, clouds form and, often, rain occurs.*

FRONTS: WHERE CURRENTS CLASH

When different air masses meet, they don't just merge together because they are different temperatures and therefore different densities. Rather than mixing, the heavy dense (cold) air slides underneath the warmer, less dense air and lifts it from underneath. The boundary between the two air masses is known as a front, named for the air behind it: as a warm front passes, warm air replaces the cooler air; when a cold front passes, cold air replaces the warmer air.

On weather maps such as you see in newspapers or on TV, the position of a front is indicated where it meets the ground. But fronts also extend upwards – sometimes into the troposphere. Fronts are around 100-200 km (60-120 miles) wide and they form along the edge of air masses. Fronts also slope, so a frontal system made up of a warm and cold front with warm air in-between is shaped a bit like a bowl – but with uneven sides. Warm fronts slope at about 0.5°-1°, a cold front slopes about 2°. Cold and warm fronts also travel at different speeds: cold fronts travel at an average speed of about 35 km/h (22 mph) while warm fronts idle along at about 24 km/h (15 mph). Because it moves more quickly, a cold front will eventually catch up with and overtake a warm front, undercut it, and lift it clear from the ground. This is known as a cold occlusion; a warm occlusion happens when the cold front rises over the warm one. When a cold front advances into warm air in a cold occlusion, it forces the warm air to rise up in a relatively steep frontal slope which triggers the formation of cumulus-type clouds. A fast-moving cold front may 'shovel up' lots of warm air and produce huge storm clouds. Each cloud quickly dissipates, but just as it does, another forms, and these create what are called squall lines, a continuous 'belt' of storms up to 1,000 km (600 miles) long.

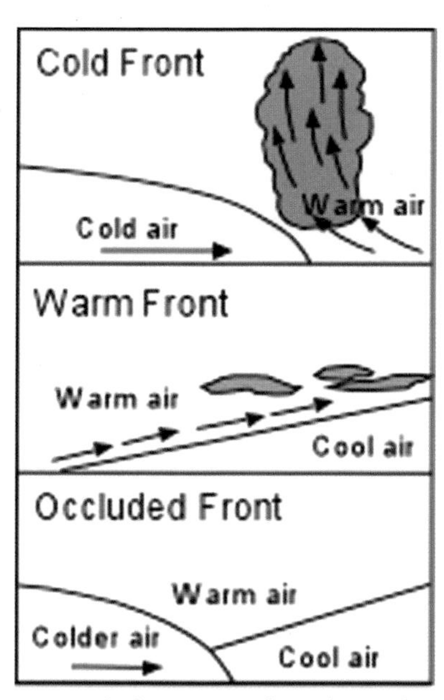

ABOVE: *A diagram to show the angles of 'slope' of different frontal sytems.*

Pressure patterns

There is a link between air masses and patterns of circulating air and these are called pressure systems. Warm air masses are associated with low-pressure systems, while cold air masses are associated with high-pressure systems. Over the course of a year, thanks to the Earth's tilt on its axis and its elliptical orbit of the Sun, the distribution of air masses and pressure systems changes. These changes have significant effects on the weather.

CYCLONES AND ANTICYCLONES

In the temperate zones of the Earth, the changeable weather we experience originates on the undulating line where the polar easterly winds clash with the prevailing westerly winds. The two currents have different temperatures and humidity levels, and when they meet they create a more-or-less permanent state of atmospheric instability. Sometimes great eddies and vortices form and move off as isolated masses of whirling air inside the general circulation system. Unlike the general circulation system, though, these eddies and vortices are maverick or episodic, that is they rise and subside. Meteorologists call them cyclones – a region where the surrounding atmospheric pressure is lower, and anticyclones – a region where surrounding air pressure is higher, and they are the mirror-image of each other.

Cyclonic winds spin in an anticlockwise direction in the northern hemisphere (clockwise in the southern hemisphere) around the centre of low pressure and meet in that centre.

LEFT: *Winds 'spin' in an anti-clockwise direction in the northern hemisphere.*

ABOVE: *'Fair weather' is the result of anticyclones.*

Anticyclonic winds spin clockwise in the northern hemisphere (anticlockwise in the southern) around a high-pressure centre and flare out from the centre. The only thing that cyclones and anticyclones have in common is that they both cover vast areas – hundreds of thousands of square kilometres.

Cyclones are the 'lows' on a weather map: they are the bringers of bad weather – clouds, storms and blizzards – but they are not the same as the extreme wind storms which are often mistakenly called cyclones! A tornado is not a cyclone! Anticyclones are the weather 'highs' and normally bring fair weather. Highs and lows account for the temperate zone's variable day-to-day weather: as they move around the Earth from west to east, cyclones and anticyclones bring clear blue skies with no rain, and rainy days accompanied by Force 8 winds!

HIGHS AND LOWS

We welcome highs because they are generally associated with lovely, calm settled weather conditions and in summer bring many sunny days. In other seasons though, highs can bring weeks – even months – of dreary, grey overcast skies. When a high hangs around and the atmosphere is too stable for too long, wintertime fog is slow to clear. This happens because, instead of decreasing with altitude as it normally would, the temperature increases – resulting it what is called a temperature inversion. Depending on the season and the temperature, an inversion layer can extend upwards a few metres to as much as 200 m (600 ft) above the ground. Acting like a saucepan lid, the inversion layer stops warm moist air – and smoke – from rising, so not only do skies stay grey for longer, impurities in the air are also trapped close to ground level and pollution levels rise.

Highs don't guarantee good weather, and fortunately,

MAIN IMAGE: *High pressure equals a stable atmosphere, which brings fair weather. But if the 'high' hangs around, there can be a temperature inversion, which then stops warm, moist air – and pollution – from rising, creating a veil of fog.*

lows don't always guarantee bad! Under certain conditions a low can bring a change for the better: if there is a sufficient interval between disturbances, the rear sky behind a cold front can sweep away fog, mist and rain and usher in extended periods of sunshine, which, now clear of impurities, feels quite warm!

UPPER-AIR DISTURBANCES AND LOW PRESSURE

The Sun is steadily beating down on the Tropics, so there is a continuous build-up of heat that keeps pumping energy, making the warm tropical air move towards the poles. As this happens, the upper-level airflow becomes unstable and begins to generate disturbances – a bit like the swirling currents in a fast-moving river as the water tries to get around rocks, and boulders. What start out as little vortices soon gather strength, grow to variable sizes and give rise to areas of low pressure. The same kind of low pressure, or cyclonic, motion can be induced by the very high mountain ranges of the Himalayas and the Rockies: both are constantly susceptible to low pressure formation. Lows, where tropical and polar air meet, are zones where energy is transferred: in the northern hemisphere, warm air from the south is pulled northwards and cold air is funnelled back down to the Tropics for reheating. Consequently, lows are vital for redistributing heat between the polar and tropical regions, keeping the heat in the atmosphere almost perfectly balanced.

MID-LATITUDE DEPRESSIONS

In between the polar and the tropical regions are the mid-latitudes whose

depressions, or lows, travel from west to east, with one low following another. These depressions are caused by Rossby waves – horizontal waves in the jet streams at the top of the front between tropical and polar air, as well as similar waves at surface level. Rossby waves are named after the Swedish scientist Carl-Gustav Rossby who worked under Vilhelm Bjerknes, and who first identified the jet streams in the 1920s (see page 70).

A crest in a Rossby wave that projects into cold air on the polar side of a jet stream produces a ridge of high pressure. A trough in the jet stream, projecting into warm air on the equatorial side of the jet stream produces low pressure. These ridges and troughs are where the winds are the strongest. At the surfaces, anticyclones generally occur downwind of the ridges, while cyclones occur downwind of troughs. A mid-latitude depression develops the characteristic arrangement of warm and cold fronts: the system begins as a straightforward boundary between warm and cold air, with the air on either side of it flowing in opposite directions. As it travels eastwards, separate fronts start to develop: the cold fronts, travelling more quickly than the warm fronts, 'catch up' and merge, form an occlusion, and then dissipate.

Tropical Cyclones: Hurricanes and Typhoons

HURRICANES AND TYPHOONS are properly known as tropical cyclones. A tropical cyclone is an extremely intense, compact – usually about 200-500 km (135-310 miles) in diameter – storm system that has formed around a core of extremely low pressure. Its vertical structure reaches from the surface of the sea to about 12 or 15 km (7 or 9 miles). Tropical cyclones develop only where a large area of the sea's surface reaches a temperature of 27°C (80.5°F): once the sea reaches this temperature, vast quantities of water are evaporated and the warm, moisture-laden air is carried upwards by a spiral of strong winds (the Coriolis effect makes the winds spiral), forming into huge clouds. As the water vapour condenses into water droplets, the release of latent heat warms the air even more, so the process repeats itself, constantly adding more 'fuel' to stoke the growing storm and making it self-sustaining. The life-span of a tropical cyclone may be as long as 30 days, or as short as a few hours.

ABOVE: *Trees in a hurricane: the powerful winds are created by differences in the atmospheric pressure of tropical cyclones.*

Because of the marked difference in atmospheric pressure, unusually powerful winds are created, often ranging from 150-250 km/h (95-160 mph). Huge thunderstorm-producing clouds spiral inwards towards the hurricane's centre – called the 'eye' – where pressure is lowest, and these swirling bands of clouds unleash torrential rain. To make things even worse, the hurricane-force winds coupled with low atmospheric pressure cause the sea to pile up near to the eye of the tropical cyclone, so there is a rise in water levels by several metres as the storm moves towards land. This storm surge not only causes flooding, but the amount of water dumped on land can result in devastating landslides as well, especially in regions where there has been significant deforestation.

Once the tropical cyclone hits land, the supply of warm, moist air is suddenly cut off and, like a car running out of fuel, the storm starts to lose its force. This is why coastal regions always bear the full impact of their fury. However, it has been known for tropical cyclones to weaken considerably as they 'hit' islands – such as those in the Caribbean – only to re-intensify once they head back out to sea!

The conditions to create a tropical cyclone occur mainly in late summer when the sea is nicely warmed, between latitudes 5° and 20°: the northern Atlantic and Pacific oceans, the Indian Ocean, the China and Arabian seas. Hurricanes take their name from the Spanish

BELOW: *Pressure is lowest at the eye of a tropical cyclone, where the sea 'poles' up.*

ABOVE: *Coastal regions bear the impact of tropical cyclones.*

MAIN IMAGE: *A hurricane approaches.*

huracan and Portuguese *huracao*, both of which are said to be derived from the original Carib name for the god of storms, urican. They sweep in from the Atlantic about 10 times a year, roughly between the months of May and September. Typhoons – from the Chinese tai fun meaning 'great wind' – generally make their appearance in August and September, but can occur in any month, and blow on average about 20 times a year in the north Pacific alone.

THE SAFFIR-SIMPSON SCALE

When the wind speed of a tropical cyclone is about 62 km/h (38 mph) or less, the storm is called a tropical depression; when winds range between 63 and 118 km/h (about 39-73 mph) it's called a tropical storm; a full hurricane is recognised if the wind speeds reach 119 km/h (74 mph) or more. The strength of tropical cyclones is measured using the Saffir-Simpson Scale (it's full title is the Saffir-Simpson Damage

THE SAFFIR-SIMPSON SCALE

CATEGORY	CENTRAL PRESSURE		WIND SPEED		STORM SURGE	
	in	hPa	km/h	mph	m	ft
1 Weak	<28.94	<980	104-133	74-95	1.2-1.5	4-5
2 Moderate	28.50-28.91	965-979	134-154	96-110	1.8-2.5	8-8
3 Strong	27.91-28.47	945-964	155-182	111-130	2.8-3.7	9-12
4 Very Strong	27.17-27.88	920-944	183-217	131-155	4.0-5.5	13-18
5 Devastating	>27.17	>920	<217	<155	<5.5	<18

** hPa = hectapascals (1hPa = 1 millibar)*

Potential Scale) which has a maximum rating of 5. The scale was originally devised to describe the potential severity of a hurricane, but was subsequently applied to all tropical cyclones and gives an assessment of both wind speeds and potential storm surge heights.

In a category 5 hurricane, where wind speeds are frequently over 217 km/h (155 mph) the force is strong enough to rip off roofs and smash wooden buildings to pieces. This incredibly destructive force is made even worse by the fact that, on either side of the hurricane's eye, the wind blows in opposite directions: as the eye passes overhead, the wind stops and then reverses, which rips apart anything that was loosened initially by the first wave of wind.

Although the wind is extremely violent, in terms of deaths related to tropical cyclone, it comes second to water. Where land is low-lying, sea water is forced inland in storm surges which can be up to 8 m (26 ft) high. Such storm surges are a hazard in the Gulf of Mexico – the greatest recorded height of a storm surge was 7.3 m (24 ft) brought by Hurricane

Camille at pass Christian, Mississippi on 7 August 1969 – but they can also strike at the USA's Atlantic coast from Florida to the Carolinas. Although deaths in the USA from hurricanes have declined in recent decades, thanks to improved forecasting and greater levels of preparation, the increasing coastal development from Texas all the way to Maine means that there are more people at risk: Florida's population has doubled to 14.6 million since 1970!

ABOVE: *Hurricane Andrew*

Storm surges also pose a considerable risk in the Ganges delta in the northern part of the Bay of Bengal, where the land is only a few metres above sea level at normal tides, and is home to millions of people. In the Bay of Bengal, the gentle slope of the sea bed increases the chances of large storm surges forming when the winds blow towards the coast and, since 1980, storm surges in this region have claimed over half a million lives.

WHAT'S IN A NAME?

Each season, hurricanes are each given a name. The names start with the first letter A, such as Hurricane Andrew, then B, then C, and so on. If a hurricane causes great damage, then, so as not to tempt fate, its name is never used again. This means we will never again have hurricanes David or Frederick (1979), Allen (1980), Alicia (1983), Elena and Gloria (1985), Gilbert and Joan (1988), Hugo (1989) Bob (1991), Andrew (1992) or Mitch (1998)!

The largest hurricane ever recorded was Hurricane Gilbert in September 1988 which reached an immense 3,500 km (2,175 miles) in diameter – about three times the average size of a Caribbean hurricane and exceeding even the biggest Pacific typhoons. Its central pressure dropped to 888 mbar, a record for the northern hemisphere, and wind

MAIN IMAGE: *Storm surges, rising sea levels caused by low pressure and coastal flooding are often more devastating then the winds themselves.*

ABOVE: *The aftermath of a hurricane.*

speeds were over 200 km/h (125 mph). Forming east of Barbados, Gilbert unleashed torrential rain on Jamaica but by the time it reached the Mexican coast, it had largely blown itself out.

Hurricane Andrew, which left a trail of damage across the southern states of the USA over five days from 23-27 August 1992, was the most expensive natural disaster to hit the USA, causing an estimated $26.5 billion worth of damage. Andrew crossed the Bahamas as a category 4 hurricane before moving on to Dade County in southern Florida where the central surface pressure dropped to 922 mbar and the maximum sustained wind (averaged over 1 minute at a height of 10 m/30 ft) was 220 km/h (135 mph), gusting to 265 km/h (140 mph). Andrew caused 15 deaths in Dade County and left 250,000 people without homes.

Andrew took four hours to cross the Florida peninsula: as it moved it lost force and weakened to a category 1 hurricane. But once it hit the open waters of the Gulf of Mexico it began to regain strength before it made its second, though less devastating, landfall at Morgan City on the coast of Louisiana. Andrew then began moving north inland, and although losing power all the time, it was still ten hours before

Andrew was downgraded to a 'mere' tropical storm!

The summer of 1988 was the first time since 1892 that the north Atlantic experienced four hurricanes, all at the same time! They were George, Ivan, Jeanne and Karl. Luckily, the last three stayed well out to sea, although there was a real threat to shipping. George, unfortunately, went on to rampage through the Caribbean, causing devastation in the Dominican Republic where virtually the entire food crop was destroyed. As Hurricane George began moving towards the Mississippi delta, 1.5 million people were evacuated from New Orleans. The city is 2m (6 ft 6 in) below sea level and relies on huge levees and drainage channels built to withstand a category 3 hurricane. George's eye made landfall at Biloxi, Mississippi with winds of 165 km/h (120 mph). After George, it all went quiet for a while.

In the Caribbean, on 21 October 1998, about 580 km (370 miles) south of Jamaica, Hurricane Mitch had formed. By early on 24 October, Mitch had

BELOW: *After causing widespread destruction on Puerto Rico, Haiti and the Dominican Republic, Hurricane Jeanne was weakened to Tropical Storm status for several days before it regained strength over the Bahamas as a Category 2 hurricane. When Jeanne made landfall in U.S. territory on September 26 it was the fourth major hurricane of the 2004 Atlantic hurricane season to strike Florida.*

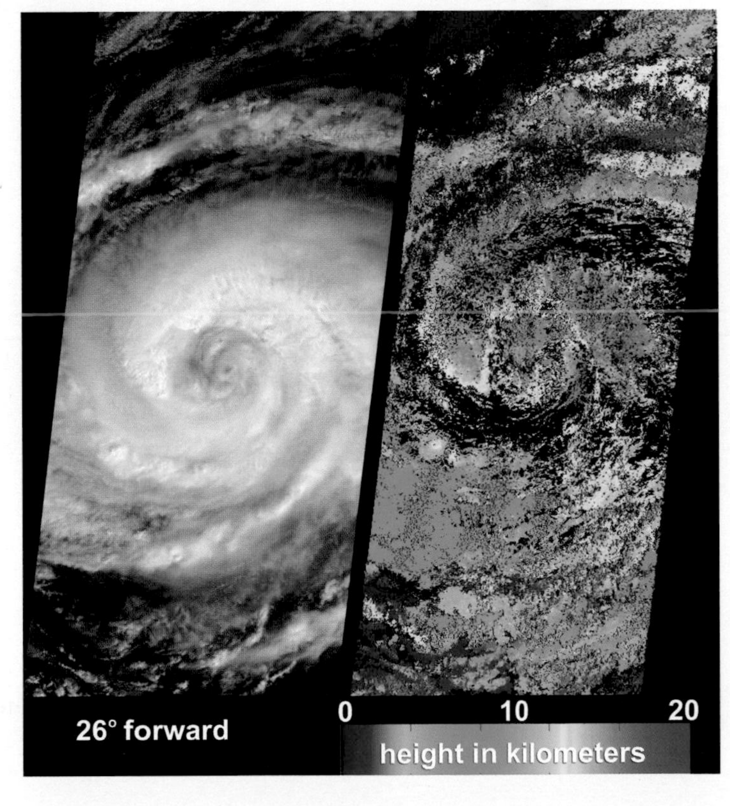

deepened and reached hurricane intensity with a surface pressure of 905 mbar and maximum sustained winds thought to be around 285 km/h (180 mph). Mitch hovered menacingly around the coast of Honduras for some time before making landfall on the morning of 29 October, then moved south across Honduras. Neighbouring Guatemala was also affected, as was Nicaragua. Although by now slow-moving, Mitch unleashed torrential rainfall: an estimated 125 mm (50 in) of rain fell in two days on Honduras. As the rainwater ran off mountainsides and poured towards the coast, it gathered tonnes of loose volcanic soil, and towns and villages were flooded and trapped in mudslides. Although the 'official' death toll was about 9,000, around the same number of people were also missing, and presumed dead, so the true number of fatalities from Hurricane Mitch will never be known.

European gales

Tropical cyclones, as their name suggests, occur in the tropical latitudes where they develop over large bodies of warm water. Sometimes similar 'disturbances' develop over the Mediterranean Sea, but Europe is too far north and the

BELOW: *The view from space of a hurricane approaching Florida.*

waters of the Mediterranean are not warm or big enough for these depressions to be classified as hurricanes. But they can, nevertheless, be pretty violent storms!

Although the Caribbean, the Pacific and the Indian ocean areas bear the full force of tropical cyclones, other parts of the world can still be affected by them. In August 1985, the 'leftovers' of Hurricane Charley intensified into a deep depression that brought gale-force winds, heavy rain, widespread flooding and severe damage to most of Great Britain. Two years later, Hurricane Floyd lifted warm air to very high levels which caused a deep depression that unleashed violent storms from 15-17 October 1987 across Britain, France, the Netherlands and southeast Norway, which suffered a violent storm surge along its southern coast.

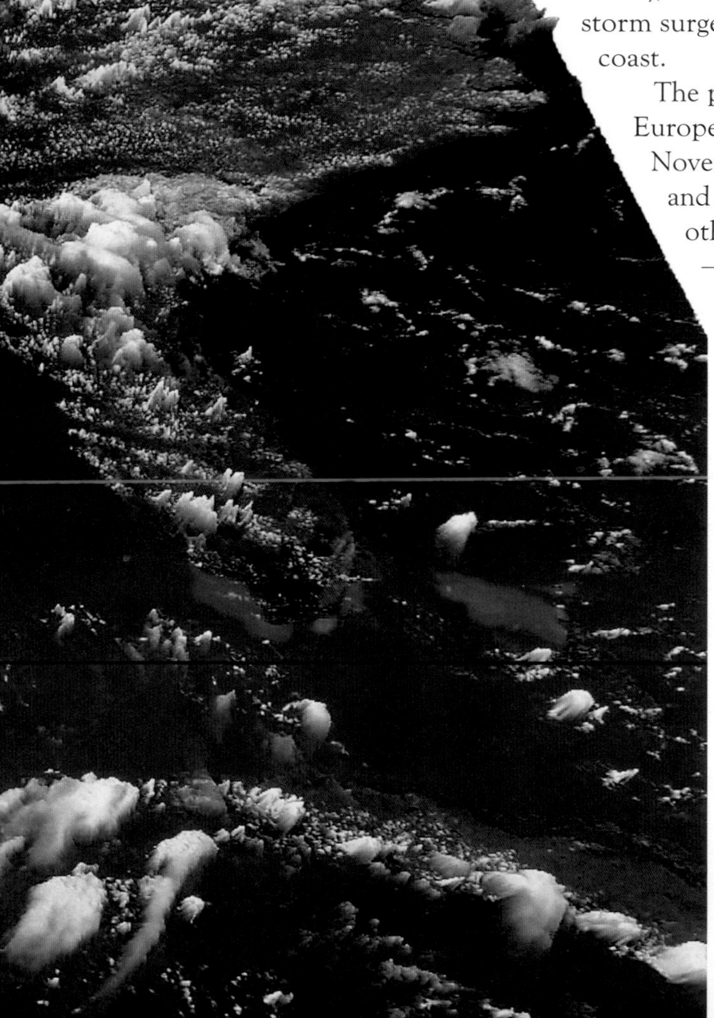

The peak season for European storms is November to February, and they are rare at other times of the year – which is why the 1985 and 1987 storms have gone down in the annals of European weather lore!

In northern Europe, gales are primarily linked with atmospheric disturbances from the west: as they develop and travel over the Atlantic Ocean, some low-pressure systems grow strong westerly winds that batter the coastlines and land masses of western Europe. In southern Europe, low-pressure systems develop over the Mediterranean and create gale force winds, along with torrential rain, before heading inland across Spain, Italy and southern France. Typically, there is a mass of cold air at high altitude, while underneath it is warm air that has been carried in from the Sahara which then picks up moisture as it crosses the Mediterranean. The low-pressure system that develops is made more unstable by orographic lifting – the lifting of the air mass up and over large geographic features (the Alps, the Pyrenees, the Cévennes and the Cantabrian mountains). Such low-pressure systems over southern Europe can produce floods and storms that range across southeastern France and as far north as the Swiss Alps.

Tornadoes: the fiercest winds on Earth

While hurricanes are the largest storms on Earth, the most destructive are tornadoes. The word tornado is a Spanish word, the past participle of *tornar*, meaning 'to turn' and from the Catalan word *tornado* meaning 'thunderstorm'. Technically, a tornado is a violently rotating column of air, normally with a small diameter, which extends down from a turbulent cloud to the ground. Tornadoes are the fiercest winds on Earth, reaching speeds of up to 480 km/h (300 mph), and they can happen anywhere! The supreme destructive power contained in a restricted area puts all life in the path of a tornado in danger: as it passes by, it sweeps the ground clear of all moveable objects. There is, however, a whole 'family' of related spinning columns of air known as whirls, devils and spouts.

MAIN IMAGE: *Tornadoes are the fiercest winds on Earth: reaching speeds of up to 480 km/h (300 mph) they can happen anywhere in the world.*

Twisters: little devils

Whirlwinds are small, local, rotating columns of air that do little damage: they are often visible lifting leaves and papers off the ground and depositing them further down a street or across a field. These small whirlwinds are often popularly called 'twisters' – a name that is used by the general public for small tornadoes, although meteorologists are not happy about this use, because 'twisters' are not tornadoes. Dust devils and sand devils are familiar sights in dry, desert country: these are spiralling columns of dust- and sand-filled air and can be several metres high. There are also snow devils in snowy regions.

Waterspouts

A waterspout is a rapidly rotating column of air that originates over water. The strongest waterspouts are 'true' tornadoes which have formed or moved over a body of water, but the majority differ from tornadoes in their origins, structure and life cycle. Furthermore, waterspouts are usually less intense, and therefore less destructive, than tornadoes, and usually lose power when they make landfall.

Waterspouts occur frequently in the Mediterranean Sea and occasionally in the Atlantic Ocean,

LEFT: *Dust devils are spiralling columns of dust-filled air and are a common sight in dry, open country.*

and there have even been incidences in the English Channel and in British coastal waters further north: it is thought that two or three waterspouts sighted near the Tay Bridge in Scotland may have contributed to its disastrous collapse on 28 December 1879 which claimed the lives of 74 people. Sailors are cautious of waterspouts, as the air pressure can cause breathing difficulties and perforated eardrums – and cause damage to boats. On 11 September 1970, a waterspout in the Gulf of Venice overturned a ferry, killing 36 of the 60 people on board.

ABOVE: *A waterspout forming, caused by a downward wind rotating in a column of air above the sea.*

The first sign that a waterspout is forming is often a dark spot or spots on the water surface, one of which usually becomes the 'primary spot'. These dark spots are variations in the surface waves, and are caused by the downward wind rotation of the air column above. In a short time a pendant forms, extending down from the clouds in the form of a visible funnel of sea vapour. This vapour forms as water vapour condenses – it is not sea water being 'sucked up'. Some sea water is trapped in the lowest part of the funnel, close to the sea's surface, called a bush. If the wind speed increases beyond about 80 km/h (50 mph), the funnel descends and throws up a ring of spray a metre or two above the sea's surface.

The funnel then begins to grow, tilt and move across the sea. At the same time, the funnel gets tighter and spins faster. Usually a waterspout will range between 300 and 700 m (about 1,000-2,300 ft) – but heights up to 1,500m (5,000 ft) have been known. In Europe, this 'mature phase' of the waterspout can last anything between 2 minutes and 20 minutes, but on average, their life span is a mere 4 minutes. Once cooler air starts to enter the funnel, the waterspout starts to decay: the spiral pattern disappears and the spout disintegrates.

THE GREATEST WIND ON EARTH

While whirlwinds, dust and sand devils are caused by the intense heating of the surface of the Earth, tornadoes are caused by the clash of warm and cold air currents at high altitudes which leads to atmospheric instability and severe turbulence. Tornadoes are most notorious in North America – especially in the 'Tornado Alley' of the Great Plains which experiences more tornadoes than anywhere else in the world. However, with the exception of the Antarctic, tornadoes can occur elsewhere in the world.

A single tornado can last from just a few seconds to an hour, but the typical duration is around five minutes. To be officially defined as a tornado, the rapidly spiralling funnel of air must be in contact with the ground. Most often the damage

ABOVE: *A single tornado can last a few seconds, but the typical duration time is around 5 minutes.*

ABOVE: *To be a true tornado, the spiralling funnel of air must be in contact with the ground.*

'path' of a tornado is about 50 m (25 ft) wide with a track of 2-4 km (1-3 miles), but sometimes the damage path can be as wide 2 km (1 m) or as narrow as 10 m (5 ft). If debris is visible – even if there is no obvious funnel cloud – a tornado is happening.

In recent years, the study of the exact mechanism that forms a tornado, known as tornadogenesis, has occupied many meteorologists. There is still, however, a great deal of mystery surrounding tornadoes and today, even when the weather conditions seem 'perfect', it is still impossible to predict their formation with absolute accuracy. What we do know about tornadoes is that they are formed by what meteorologists call 'super cell storms', that they are always linked to a parent cumulonimbus cloud (see page 124), and that during their 'life' they undergo certain changes.

SUPER CELL STORMS

Tornadoes are usually associated with thunderstorms because they require a moist air stream that is warm and that usually comes from a southerly direction: most tornadoes in the USA occur when air temperatures are above 18°C (64°F) and generally, tornadoes move from the southwest to the northeast. They rarely move westwards but they have been known to move around, change direction, zig-zag, stand still, and travel in a complete circle! In the northern hemisphere, tornadoes usually spin in an anticlockwise direction.

A super cell storm is an extremely violent and persistent storm which is characterised by a large rotating up-draught (called a mesocyclone or tornado cyclone) that extends high into the top of cumulonimbus cloud. Mesocyclones are about 16 km (10 miles) in diameter and contain 'mini' warm and cold currents that interact to form the funnel of the tornado. When the wind alters speed or direction near the top of a cumulonimbus cloud, it tilts the warm up-draughts. Cold down-draughts then fall to the side of the rising air without cooling it. This allows the cloud to grow much

LEFT: *The damage 'path' of a tornado is generally around 50 m (25 ft) wide, but its track (the distnace it travels) can be a long as 2-4 km (1-3 miles).*

bigger and last much longer than an ordinary storm cloud.

The funnel cloud

The warm, rising air begins to rotate slowly, starting near the top of the cloud. As it extends downwards, its rotation gets narrower and it starts to spin faster. The spinning air continues downwards, forming the distinguishing funnel beneath the cloud. Air drawn into the funnel enters an area of much lower pressure, so it expands, cools and gives up its moisture, the exchange of latent energy fuelling the process further. The visible funnel is therefore formed by condensing water vapour which is the result of the lower pressure in the whirl. But it can contain debris drawn into it as its size and strength increases. A tornado funnel can also take several forms, ranging from a grey-white, thin, rope-like, writhing column, to a menacing thick mass of black. It is also possible that several funnels develop in a tornado 'system': small whirls or vortices may continually form and fade while they whirl around the central core of the main tornado circulation.

Touch down

When the rotating funnel of air reaches the ground, by definition the storm (or 'disturbance' as meteorologists call it) does become a tornado. During a tornado's 'mature phase' the funnel reaches it greatest width and is almost always nearly vertical. Most of the time, it's touching the ground – although tornadoes can play hopscotch along the way. It's when a tornado is mature that it causes the most damage: the worst American tornado for fatalities was the Tri-State outbreak on 18 March 1925 when 689 people lost their lives across Missouri, Illinois and Indiana.

Oklahoma, 1999

The worst tornado outbreak since 1925 hit Oklahoma in May 1999. Even though the state is frequently visited by tornadoes, the events of 3 May were extreme and affected a heavily populated area. No fewer than 50 tornadoes ran across central Oklahoma, but the killer was a massive F5 tornado with a record wind speed of 512 km/h (318 mph): 40 people died in and to the southwest of Oklahoma City. There were also tornadoes on the same day in north Texas, eastern Oklahoma and south central Kansas: five people died in Wichita.

Although not on the same scale as American tornadoes, the largest outbreak of tornadoes in Europe occurred in Britain on 21 November 1981 when 105 tornadoes moved across the country from Wales to

RIGHT: *A super cell thunderstorm that spawned a tornado, Kansas, USA.*

Wales to Humberside (in the north), Essex (in the southeast) and Norfolk (in the east)! No-one was killed, but European tornadoes can be devastating: on 9 June 1984, 400 were killed and 213 injured when a T10 tornado hit the towns of Belaynitsky, Ivanova and Balino in western Russia.

CLASSIFYING TORNADOES

For some years meteorologists and researchers have used the Fujita-Pearson Tornado Intensity Scale (named after its creators) to rate the intensity of tornadoes. The scale provides ratings in three areas: force, or wind speed (F); path length (PL), and path width (PW). The ratings are in fact three separate scales: an F1 classification does not necessarily mean that the PL and PW are also class 1.

The force (F) represents the maximum wind speed in each storm. But because the ferocity of tornadoes rips wind instruments apart, scientists instead rely on photographs of storm damage to calculate the force or wind speed that must have been present. However, assessing wind speed from damage can be problematic – especially where tornadoes cross large areas of open countryside. In 1972 the British-based Tornado and Storm Research Organisation (TORRO) devised a 10-point scale directly related to wind speed with increasingly accurate measurements of wind speed obtained from Doppler radar systems.

In Britain and Europe, the majority of tornadoes (about 94%) are rated T0 to T3 – approximately equivalent to an F0 or F1 tornado on the Fujita-Pearson Scale. Neither scale, however, has been officially adopted by the World Meteorological Organisation (WMO).

FUJITA-PEARSON SCALE

SCALE	FORCE		CATEGORY	PL		PW		DAMAGE
	km/h	mph		km	miles	miles/km	yd/miles	
F0	0-116	0-72	Weak	0-1.6	0-1	0-16m	0-17 yd	Light
F1	117-180	73-112	Weak	1.6-5	1-3.1	16-50m	18-55 yd	Moderate
F2	182-253	113-157	Strong	5.1-15.9	3.2-9.9	51-160m	56-175 yd	Considerable
F3	254-332	158-206	Strong	16-50	10-31	161-508m	176-556 yd	Severe
F4	333-418	207-260	Violent	51-159	32-99	0.54-1.4km	0.34-0.9 mi	Devastating
F5	420-496	261-308	Violent	161-597	100-315	1.5-5km	1-3.1 mi	Incredible

TORRO TORNADO INTENSITY SCALE

SCALE	WIND SPEED		EXPECTED DAMAGE
	m/s	km/h	
T0	17-24	61-86	Light
T1	25-32	90-115	Mild
T2	33-41	119-148	Moderate
T3	42-51	151-184	Strong
T4	52-61	187-220	Severe
T5	62-72	223-259	Intense
T6	73-83	263-299	Moderately Devastating
T7	84-95	302-342	Strongly Devastating
T8	96-107	346-385	Severely Devastating
T9	108-120	389-432	Intensely Devastating
T10	above 121	above 436	Super

CLOUDS

THE WEATHER STATIONS IN THE SKY

CLOUDS APPEAR IN MANY STRANGE and beautiful shapes and, at first, it seems that no two clouds are ever alike. Clouds are very useful to humans on the ground for, unlike the wind, they are visible: clouds tell us what is going on in the different layers of the atmosphere and give us useful clues about what may happen in the weather in the hours and days to come.

THE VISIBILITY OF CLOUDS means that, like the Sun and the Moon, they became important weather 'signs' for our ancestors. By the 4th century BC many Greek philosophers and naturalists were busy writing about the weather and associated optical phenomena: Aristotle's *Meteorologica* commented on clouds as well as rain, dew, snow, hail, storms and rainbows. His disciple, Theophrastus, who lived at Eresus on the island of Lesbos, not only wrote a treatise on the winds but also made observations – albeit rather generalised ones – about clouds: 'If in fair weather,' he wrote, 'a thin cloud appears stretched in length and feathery, the winter will not end yet.'

Thin, thick, fluffy, tufts, billows, plumes, cotton wool, mashed potato, cauliflower, mare's tails and mackerel, are all common ways of describing different types of clouds in layman's terms. What was needed was a scientific classification system and that didn't happen until the early 19th century!

In 1802, the French naturalist Chevalier de Lemarck published a list of cloud types with five main divisions: hazy, massed, dappled, broom-like and grouped. Lemarck's classification did not meet with widespread approval as his treatise also included references to the Moon (which he believed influenced the weather).

Howard's Cloud Classification System

THE FOLLOWING YEAR, Luke Howard, English pharmacist and weather observer, wrote an essay titled 'On the Modification of Clouds' which, with its use of internationally understandable Latin names, laid the groundwork for the classification system still in use today. Howard's essay was enthusiastically received by the international scientific community – as well as by the German poet Goethe who dedicated four nature poems to Howard.

Howard's three main 'modifications', or classifications, were named cirrus ('curl', 'tuft' or 'wisp') for high, wispy, fibrous, parallel strands of clouds; cumulus ('heap' or 'mass') for the heaps of clouds with dome-shaped tops that grow up from their flat bases; and stratus ('layer' or 'spread out') for the low, horizontal, extensive and continuous sheets of cloud that grow

FAR LEFT TO BELOW: *Cirrus, cumulus, stratus. Howard's classification of clouds was made in 1803.*

from top to bottom. The division of clouds into three categories is related not only to their composition, but also to the way in which they are formed.

A fourth category was also added – nimbus ('rain') for rain clouds. Howard assumed that clouds could change from one type to another so the various combinations of the three categories plus the nimbus type of clouds became the basis for the ten main cloud genera or 'families': cirrus, cirrocumulus, cirrostratus, altocumulus, altostratus, nimbostratus, stratus, stratocumulus, cumulus and cumulonimbus.

INTERNATIONAL ACCEPTANCE

In 1887, the English meteorologist Ralph Abercromby travelled round the world to make sure that the clouds in categories suggested by Howard looked the same everywhere: they did. The 1891 International Meteorological Conference in Munich, Germany, recommended that Howard's classification system be adopted by all the weather services.

In 1896, The International Cloud Atlas, the first comprehensive collection of cloud photographs and descriptions was published. It was revised (in colour) in 1956 by the Geneva-based World Meteorological Organisation and included new information made available by advances in aviation and

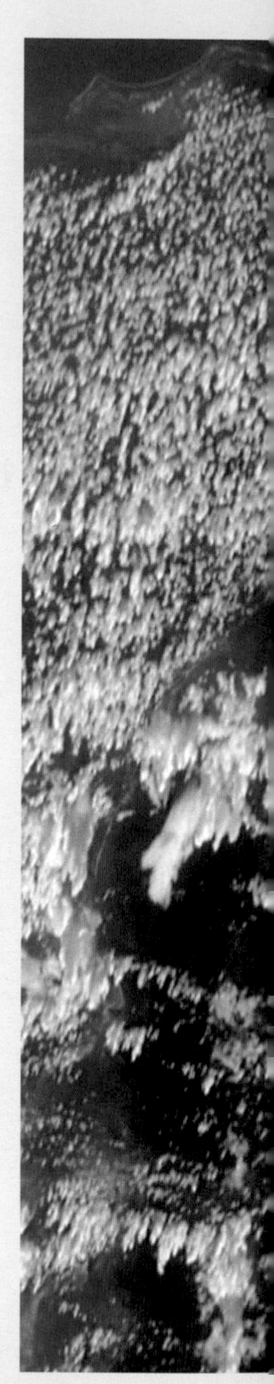

RIGHT: *Despite their delicate appearance, thin, feathery clouds of ice crystals, called cirrus, may contribute to global warming. Some scientists believe cirrus is quite common, but it is notoriously difficult to observe -- even from satellites, which offer our only means of monitoring such clouds over the entire planet.*

MISR
MAS
9 July 2002

technologies. In 1987 the two-volume atlas was further revised and expanded and now includes the 10 cloud genera along with 14 cloud species (to describe the differences in cloud shapes and structure), 9 cloud varieties (describing cloud transparency and the arrangement of cloud elements), 3 forms of accessory clouds and, 6 supplementary features that particular genera or species may adopt. Today, *The International Cloud Atlas* remains the standard reference for meteorological services in identifying cloud conditions for the daily weather map.

CLASSIFICATION ACCORDING TO HEIGHT

Meteorologists also classify clouds according to their height above ground – high, middle or low altitude. In all cases, the height is taken to be that of the base of the cloud. Heights are given in metres, although the international aviation industry continues to quote altitudes in feet, so equivalents are also given. It is important to remember that cloud heights vary considerably depending on latitude (the defining altitudes are lower in the polar regions and higher in the Tropics) and time of the year, so the figures given should be considered as approximate heights.

Low-altitude clouds have their bases below 2,000 m (6,000 ft). Between 2,000 and 5,500 m (6,000 and 18,000 ft) are the middle-altitude clouds: they usually have the prefix 'alto' meaning 'middle' in meteorology, and they form the main portion of active weather systems. High-altitude clouds generally have bases above 5,500 m (18,000 ft) and they exist in the atmosphere where temperatures are well below freezing and are composed of ice crystals. Cumulonimbus clouds, however, can extend through more than one level.

The various names of clouds have standardised two or three letter abbreviations which makes it a lot easier and faster to write down cloud observations. The full classification scheme follows on the next pages.

CLOUD GENUS

THERE ARE 10 BASIC cloud types which, between them, represent a number of overall characteristics. Their names are abbreviated to two letters.

ALTOCUMULUS Ac

Heaps or rolls of clouds, showing distinct shading and with clear gaps between them, in a layer at middle levels. These clouds can be found worldwide, and all year round. They often form at night and are associated with cold fronts, especially in temperate climates. Composed mainly of water droplets.

ALTOSTRATUS As

Sheets of featureless white or grey cloud at middle levels. These clouds can be found worldwide but are most common in mid-latitudes and are associated with warm fronts, especially in temperate climates. They contain ice crystals near the top, but water droplets lower down.

CIRROCUMULUS Cc

Tiny heaps of cloud with no shading, with clear gaps, in a layer at high levels. Occur worldwide and all year round and are associated with cold fronts, especially in temperate climates. Consist entirely of ice crystals. Small vertical air movements can produce a 'mackerel sky' – indicating approach of strong surface level winds.

CIRROSTRATUS Cs

Essentially featureless sheet of thin cloud at high level. Worldwide distribution and occur all year round. Associated with approaching warm fronts. Consists of ice crystals. When the Sun shines on high-level Cs at dawn or dusk, it sometimes makes the cloud appear pink, producing spectacular sunrises and sunsets. At other times it appears as a white veil that makes the sky look milky.

CIRRUS Ci

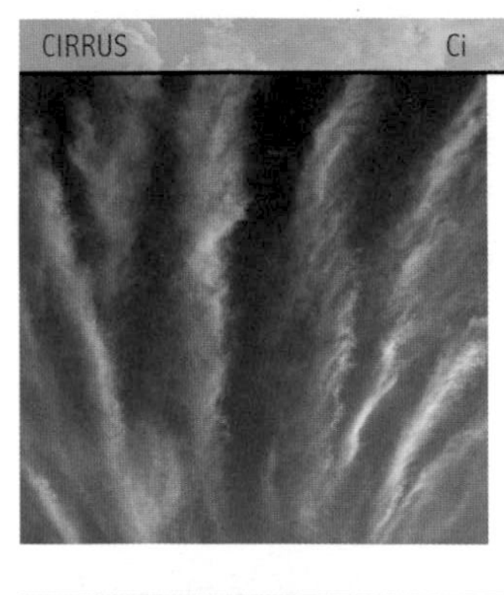

Fibrous wisps at high level. Found worldwide, all year round. Consist of ice crystals which grow until they reach max. size of 0.05 mm (1/500th in) across, when they become heavy and sometimes drift downwards into a wind flowing in a different direction which stretches the cloud into long, often curving tails – mare's tails, or cirrus uncinus. Their appearance usually indicates that the ground level wind will soon strengthen.

CUMULONIMBUS Cb

Large towering cloud extending to great heights – sometimes into the troposphere and beyond – with ragged base and heavy precipitation. Found worldwide except for Antarctica; some types especially common in tropical regions. They occur all year round, but rarely survive for more than one hour. Associated with tropical cyclones, thunderstorms, tornadoes, torrential rain, hailstorms and snow. Composed of liquid cloud droplets in lower regions and ice crystals near the top.

CUMULUS Cu

Rounded heaps of clouds at low level. Found worldwide, common in humid regions. All year round, but more common in summer. Can vary greatly in size (up to 10 km/6 m across and 20 km/12 miles high); usually separate from each other with blue sky between. Bases mark level (2,000 m/6,500 ft) at which water vapour condenses so all cumulus clouds in one area are at the same height.

NIMBOSTRATUS NS

Dark, grey cloud at middle level, often extending towards the surface, and giving prolonged precipitation. Found all year round, worldwide except in Antarctica. If cloud temperature is above -10°C (14°F) throughout, the cloud will consist mainly of water; from -10°C to -20°C, (14°F to -4°F) it will be a mixture of water and ice crystals; below -20°C (-4°F) it will consist mainly of ice. This mixture favours the formation of larger ice crystals that fall as snow, or melt, and fall as rain.

STRATOCUMULUS Sc

Heaps or rolls of clouds, with distinct gaps and heavy shading at low level. Found worldwide, all year round. Composed entirely of water droplets. Form when rising air meets a layer of warmer air and is flattened against it. This produces an overcast sky. At dusk and dawn, sunlight shining through gaps may illuminate dust particles producing rays that converge on to the clouds from below.

STRATUS St

Essentially featureless, grey layer cloud at low level. Found worldwide but more common near mountains and coasts. Occurs year round but more common in winter. Consists entirely of liquid droplets. Cloud base is usually lower than 2,000 m (6,500 ft) and if it is at ground level, then stratus occurs as fog. (Hill fog has a base lower than the hilltop it covers.) Stratus also forms in valleys. As there is no vertical air movement within stratus, it cannot produce rain, snow or hail, but drizzle or snow grains sometimes fall. Forms overnight in fine weather, especially over water and dissipates quickly in the morning as temperatures rise and cloud droplets evaporate. Frontal stratus forms along warm fronts where warm, stable air is being slowly raised up by cooler air.

CLOUD SPECIES

THERE ARE FOURTEEN TERMS USED to describe the shape and structure of clouds. Each has a name abbreviated to three letters.

SPECIES	ABBREVIATION	DESCRIPTION	GENERA
Calvus	cal	Tops of rising cells lose their hard appearance and become smooth	Cb
Capillatus	cap	Tops of rising cells become distinctly fibrous or striated: obvious cirrus may appear	Cb
Castellanus	cas	Distinct turrets rising from an extended base or line of cloud	Sc, Ac, Cc, Ci
Congestus	con	Great vertical extent; obviously growing vigorously, with hard 'cauliflower-like' tops	Cu
Fibratus	fib	Fibrous appearance, normally straight or uniformly curved; no distinct hooks	Ci, Cs
Floccus	flo	Individual tufts of clouds, with ragged bases, sometimes with distinct virga	Ac,Cc, Ci
Fractus	fra	Broken cloud with ragged edges and bases	Cu, St
Humilus	hum	Cloud of restricted vertical extent; length much greater than height	Cu
Lenticularis	len	Lens or almond-shaped clouds, stationary in the sky	Sc, Ac, Cc
Mediocris	med	Cloud of moderate vertical extent growing upwards	Cu
Nebulosus	neb	Featureless sheet of cloud with no structure	St, Cs
Spissatus	spi	Dense cloud, appearing grey when viewed towards the Sun	Ci
Stratiformis	str	Cloud in an extensive sheet or layer	Sc, Ac, Cc
Uncinus	unc	Distinctly hooked, often without a visible generating head	Ci

Cloud Varieties

THERE ARE NINE TERMS, with two-letter abbreviations, that describe cloud transparency and the arrangement of cloud elements. Any given cloud may show the characteristics of more than one variety and often several varieties may be present at the same time.

RIGHT: *Perlucidus*

VARIETY	ABBREVIATION	DESCRIPTION	GENERA
Duplicatus	du	Two or more layers	Sc, Ac, As, Cc, Cs
Intortus	in	Tangled, or irregularly curved	Ci
Lacunosus	la	Thin cloud with regularly spaced holes, appearing like a net	Ac,Cc,Sc
Opacus	op	Thick cloud that completely hides the Sun or Moon	St, Sc, Ac, As
Perlucidus	pe	Extensive layer with gaps, through which blue sky, the Sun or Moon, are visible	Sc, Ac
Radiatus	ra	Appearing to radiate from one point in the sky	Cu, Sc, Ac, As, Ci
Translucidos	tr	Translucent cloud through which the position of the Sun or Moon is visible	St, Sc, Ac, As
Undulatus	un	Layer or patch of cloud with distinct undulations	St, Sc, Ac, As, Cc, Cs
Vertebratus	ve	Lines of clouds looking like ribs, vertebrae or fish bones	Ci

Accessory Clouds

There are three forms of clouds, with three-letter abbreviations, that occur only in conjunction with one of the 10 main genera.

NAME	ABBREVIATION	DESCRIPTION	GENERA
Pannus	pan	Ragged shreds of cloud beneath main cloud mass	Cu, Cb, As, Ns
Pileus	pil	Hood or cap of cloud above rising cell	Cu, Cb
Velum	vel	Thin extensive sheet of cloud, through which the most vigorous cells may penetrate	Cu, Cb

SUPPLEMENTARY FEATURES

There are six forms, some common, others quite rare, that particular cloud genera or species may adopt. These names have a three-letter abbreviation.

FEATURE	ABBREVIATION	DESCRIPTION	GENERA
Arcus	arc	Arch or roll of cloud	Cb, Cu
Incus	inc	Anvil cloud	Cb
Mamma	mam	Bulges or pouches under higher cloud (they are named for the latin word for 'udder'!)	Cb, Ci, Cc, Ac, As, Sc
Praecipitatio	pre	Precipitation that reaches the surface	Cb, Cu, Ns
Tuba	tub	Funnel cloud of any type	Cb, Cu
Virga	vir	Fallstreaks: trails of precipitation that do not reach the surface	Ac, As, Cc, Cb, Cu, Ns, Sc, (Ci)

Two additional terms are also sometimes used: the suffixes '-genitus' and '-mutatus' which are abbreviated to 'gen' and 'mut'. These suffixes are added to a particular cloud genus to indicate the type from which a currently observed cloud is derived.

When 'gen' is added, the implication is that the 'parent' cloud is still present: for example, altocumulus cumulogenitus (Ac cugen) indicates that the dominant cloud is altocumulus (Ac) which has formed from cumulus and substantial amounts of it are still evident (cugen).

When 'mut' is added, the implication is that the parent cloud has been altered: stratus stratocumulomutatus (St scmut) indicates the presence of stratus (St) which was originally derived from stratocumulus cloud.

The scheme originally devised by Howard makes it possible to describe clouds in great detail, but it must be remembered that these descriptions are based solely on a cloud's appearance.

What the Clouds Tell Us

The actual way in which clouds are formed and their subsequent changes, are often of much greater significance than a cloud's appearance for understanding and forecasting changes in the weather.

Howard's three main divisions of clouds into cirrus, cumulus and stratus allow us to place clouds into one, or more, of the broad categories; some of the ten cloud genera, appear in more than one group because they have a combination of characteristics:

Cumuliform clouds which have more or less pronounced heads or turrets on their upper surfaces. Clouds of the following genera are in this group: cumulus (Cu), cumulonimbus (Cb), stratocumulus (Sc), altocumulus (Ac) and cirrocumulus (Cc). These clouds

LEFT: *Rounded heaps of cumulus clouds are most common in summer.*

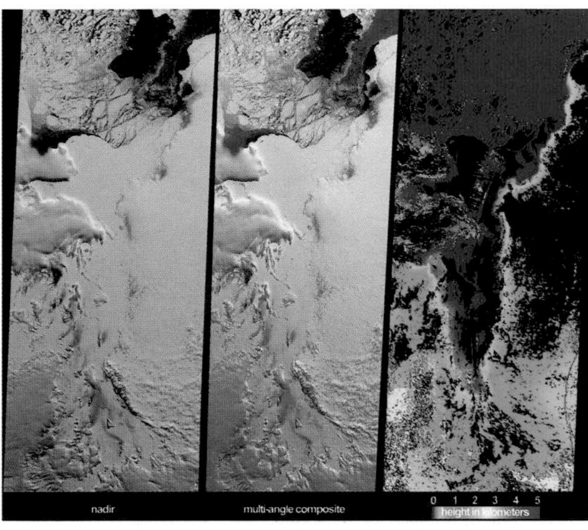

ABOVE: *These views from the Multi-angle Imaging SpectroRadiometer (MISR) illustrate ice surface textures and cloud-top heights over the Amery Ice Shelf/Lambert Glacier system in East Antarctica on October 25, 2002.*
Clouds exhibit both forward and backward-scattering properties in the middle panel and thus appear purple, in distinct contrast with the underlying ice and snow. An additional multi-angular technique for differentiating clouds from ice is shown in the right-hand panel, which is a stereoscopically derived height field retrieved using automated pattern recognition involving data from multiple MISR cameras. Areas exhibiting insufficient spatial contrast for stereoscopic retrieval are shown in dark gray. Clouds are apparent as a result of their heights above the surface terrain. Polar clouds are an important factor in weather and climate. Inadequate characterization of cloud properties is currently responsible for large uncertainties in climate prediction models. Identification of polar clouds, mapping of their distributions, and retrieval of their heights provide information that will help to reduce this uncertainty.

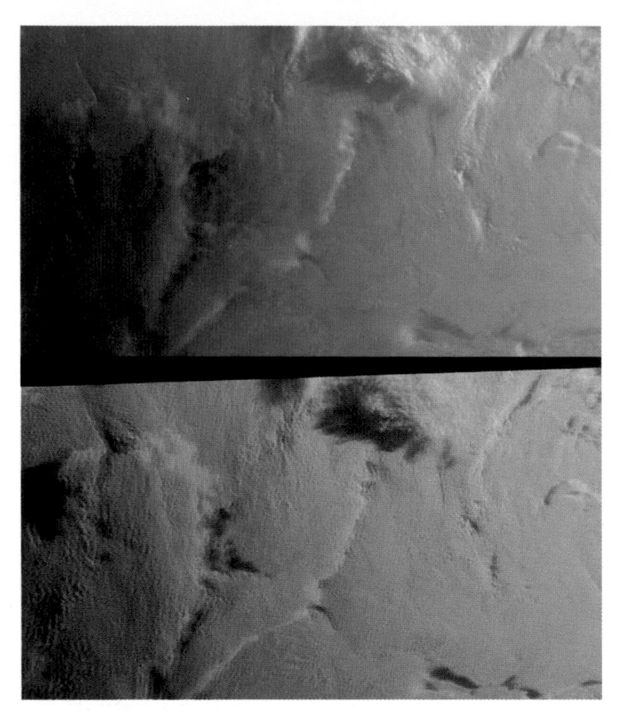

ABOVE: *Stratus clouds are common in the Arctic during the summer months, and are important modulators of the arctic climate. This image pair from the Multi-angle Imaging SpectroRadiometer (MISR) was acquired on August 23, 2000, and shows a region of stratified clouds situated near the boundary of the permanent polar ice pack to the north of the Chukchi and East Siberian Seas. At top is a natural-colour view captured by MISR's vertical-viewing (nadir) camera. At bottom, a stereo anaglyph enables observation of multiple cloud layers.*

indicate that convection is occurring in the cloud and that there is instability present in the atmosphere.

Stratocumulus (Sc), altocumulus (Ac) and cirrocumulus (Cc) clouds have features of cumuliform clouds with convection taking place inside them, but they also frequently occur in distinct layers and can therefore be placed in an intermediary position between the cumuliform and stratiform groups.

Stratiform clouds are layer clouds with generally smooth upper surfaces – although the top surface may be impossible to see when the clouds form a blanket across a large part of the sky. Clouds of the following genera are in this group: stratus (St), nimbostratus (Ns), altostratus (As) and cirrostratus (Cs). The presence of these clouds indicates that the atmosphere is stable and without vertical currents.

Cirriform clouds include clouds of the following genera: cirrus (Ci), cirrocumulus (Cc) and cirrostratus (Cs). These clouds consist mainly of ice crystals – although cirrocumulus (Cc) also contains super-cooled water droplets. Ci and Cs clouds often have distinct striations that appear like hair or threads – or 'mare's tails' which are caused by trails of ice crystals, known as virga (vir). The appearance of these clouds often warn of the approach of a warm front marking the advance of a snow storm.

HOW MUCH CLOUD?

By (almost!) international agreement, cloud 'amounts' are reported as eighths, or oktas, of the sky covered either by a specific type of cloud (which is called partial cloud amount) or by any type of cloud (referred to as total cloud amount). The amount of cloud is estimated to the nearest okta on a scale of 0 to 8: a clear sky is 0 while a completely covered sky is 8. A 'cloudy day' meteorologically, is when the average amount of cloud at the standard times of observation (with clouds, surface observations are normally carried out every hour) is greater than 6 oktas. In North America however, a scale of tenths is used instead of eighths, so a cloudy day in the USA is a day when overall cloudiness exceeds 7/10ths.

BELOW: *Wispy cirrus clouds at sunset: their appearance usually indicates that the ground level wind will soon strengthen.*

How Clouds are Formed

Clouds, mist and fog are formed when air, which always contains some moisture, falls below the dew point – the temperature at which the invisible water vapour contained in the air condenses into a collection of tiny water droplets. Meteorologists deduce this temperature from what are called dry and wet bulb temperatures (see page 226). The collection of water droplets is what we see from the ground as a cloud. If the droplets continue to acquire moisture and grow large enough, they will fall out of the cloud as rain. The height at which this occurs depends on the stability of the air and the amount of moisture present. A typical cloud droplet or ice crystal is about 0.01 mm (0.0004 in) across. The clouds formed at high altitude – the cirriform clouds, Ci, Cc, and Cs clouds – contain only ice crystals, while lower-altitude warm clouds – the cumuliform clouds – contain only water droplets, and mixed clouds such as nimbostratus (Ns) and altostratus (As) contain both.

Cloud formation can happen in two ways: warm air can rise, cool and expand, or warm air can come into contact with a cold surface.

There are various mechanisms which force warm air to rise, but the three most important ones are by convection, by orthographic lifting, or by lifting at a frontal system (see below).

WARM AIR LIFTED BY CONVECTION

Clouds, such as the cumuliform clouds, are 'created' by convection, when warm 'bubbles' of air known as thermals rise from a heated surface (normally the ground). In the individual thermals there is circulation: warm air is rising in the centre, while cooler air descends on the outside. Early in the day, thermals mix with the surrounding air and 'die' away, but later in the day, as temperatures rise, the thermals rise higher, the air expands (because air pressure decreases with altitude) and eventually cools sufficiently to reach its dew point.

All the cumuliform clouds – from the smallest tufts of cumulo fractus (Cu fra) to the enormous cumulonimbus (Cb) – are formed in this way. Convection is therefore a hugely important process in weather 'events': cumuliform clouds tell us that the air is unstable, at least in the lower levels of the atmosphere.

If the rising air meets what is called an inversion, a region of stability, the cumulus may spread out sideways into a layer of stratocumulus (Sc) or altocumulus (Ac) unless the convection is strong enough to force a way through the stable layer and continue its upward growth. Convection can also occur when the top of a cloud is cooled by radiating its heat out into space – something that often happens in a layer of stable stratiform clouds. The cooler air starts to sink and breaks up the layer into individual, shallow convective cells. A similar process also occurs in the formation of mamma (the pendulous 'udders') which may be found underneath cumulonimbus clouds.

ORTHOGRAPHIC LIFTING

Clouds can also be formed when warm air is 'lifted' within a depression, or more often, by the slope of a hill or mountain. This process is called orthographic lifting. If the wind meets a barrier, such as a range of mountains, the air is forced to rise up the slope of the hill or mountain; as the air current rises, it expands and cools. If it reaches its dew point, a fog forms. If this fog is above the surface of the Earth, it forms a cloud. The warm, moist westerly winds that blow across the Atlantic Ocean to Europe often form clouds and rain as they reach the Welsh and Scottish mountains.

The exact type of cloud formed by orthographic lifting once again depends on whether the air is stable or unstable. If the air is stable, it will tend to sink back to its original level once it passes the top of the obstacle – the peak of the hill or mountain. In this instance, the cloud will be stratiform in nature: stratus (St), nimbostratus (Ns), altostratus (As) or even cirrostratus (Cs) and it will lie only over the higher ground and not extend downwind (on the other side of the mountain range).

If the air is unstable (or if it becomes unstable when it is forced to rise upwards), the clouds will be cumuliform in nature, in particular, cumulus (Cu) and cumulonimbus (Cb). Most of the time, the wind carries these clouds away and they are replaced by new ones. However,

BACKGROUND IMAGE: *Cumulonimbus storm clouds.*

LEFT: *Orthographic lifting means that the peaks of tall mountains are often shrouded in cloud.*

sometimes, the clouds will 'hang around' over the high ground where cumulonimbus in particular will produce long and heavy rainfall.

In general, any rainfall induced by the high ground will be heaviest on the windward slopes – the sides where the warm, moist air is being lifted up – and on the peak. On the leeward side, the 'far side' of the mountain range or hills where the now cool air descends back down the slope, is often a 'rain shadow' with less rainfall.

LIFTING AT A FRONTAL SYSTEM

Warm air also gets lifted when cool air undercuts it at a weather front. This is known as frontal lifting and commonly occurs at a warm front. In broad terms, a warm front means warmer, moist air is advancing, and causes a 'veering' of the wind direction. This means that the wind shifts in a clockwise direction over the space of an hour or two from southeasterly to southwesterly in the northern hemisphere. In the southern hemisphere, the wind will shift from northeasterly to southwesterly.

High above the Earth's surface, and sometimes up to 700 km (435 miles) ahead of the front, the first cirrus (Ci) cloud will appear. As the warm front approaches a point on the Earth's surface, the base of the clouds that have been produced by the 'overrunning' warm, moist air gradually begin to lower. The clouds change from cirrus to cirrostratus (Cs) or cirrocumulus (Cc), followed by altostratus (As) and altocumulus (Ac), and then start thickening into nimbostratus (Ns), the 'rain clouds' whose precipitation will fall to Earth quite steadily. Sometimes, some of the rain falling through the really damp bases of nimbostratus clouds evaporates, which cools the air slightly. This leads to condensation in the moist air which creates 'scud' or 'fractostratus' clouds – those low, ragged clouds that fly across the sky when it's raining hard and the wind is blowing quite strongly!

In a cold front, it is the heavier, colder air that is advancing and it undercuts the warm air and lifts it up. Sometimes the clouds that are produced are similar to those in a warm front, especially in winter at middle latitudes. But the cold air 'on the move' over the relatively warm Earth's surface results in unstable conditions, which produce vigorous convection in the form of cumulonimbus (Cb), and heavy, but usually brief, showers of rain and thunderstorms. In general, then, convection clouds produce showers and brief storms, while orthographic and frontal clouds produce more persistent rain and drizzle.

NEXT PAGE: *Cumulus clouds may herald the formation of a super cell storm, which can give rise to tornadoes.*

Cloud Patterns

If you sit and look at the sky for long enough you will see that clouds form quite regular, and often very beautiful, patterns. These patterns include banners, waves, billows, cloud streets, fallstreak holes, pyrocumulus, and the cloud trails left by aircraft as they criss-cross the sky, called contrails and distrails.

BANNER CLOUDS

A banner cloud, sometimes also called a pennant cloud, is a form of orthographic cloud. These clouds are very distinctive 'plumes' that hang like flags or banners behind isolated mountain peaks and are often the only clouds in the sky. Some of the most famous banner clouds are those that 'fly' from the Rock of Gibraltar, from the Matterhorn in the Alps on the Swiss-Italian border at Zermatt, and the appropriately huge banner that flies from Mount Everest in the Himalayas. For a banner cloud to form, the wind needs to be blowing quite strongly from the cooler side of the mountain face (the face that is in shadow) to the warmer, sunnier face. The wind creates an increase of pressure on the windward side, and a corresponding decrease in pressure occurs on the leeward side, immediately behind the peak. Eddies develop in the region behind the peak which curl up over the top, dragging warm air upwards with them towards the summit of the mountain. The rising air expands, cools, and reaches its dew point, causing the water vapour to condense and form cloud droplets. The banner hangs from the peak throughout most of the day – unless there is a major shift in wind direction – and disappears at night when the warming process ends.

WAVE CLOUDS

Lenticular clouds

A second form of orthographic cloud are wave clouds. These occur when air is forced over a mountain barrier to create a series of waves that stretch downwind. Rather than simply going over the top of the hill or mountain and then sinking down the other side, the wave motion initiated by the peak continues. Sometimes there aren't any clouds, but the invisible waves, or turbulence, in the air can affect aircraft. Frequently though, when there are moist layers of air present, very distinctive clouds form in the crests of the waves. These are lenticular, lens- or almond-shaped, clouds that hang stationary in the sky, often for long periods, even though the wind speed and wind direction remain constant. The clouds appear to be stationary but

LEFT: *Lenticular, or 'almond'-shaped clouds are orthographic clouds.*

the wind is, in fact, flowing through them. Wave clouds may occur as stratocumulus lenticularis (Sc len) altocumulus lenticularis (Ac len) or cirrocumulus lenticularis (Cc len) and are common in mountainous areas. Although classed as cumuliform clouds, convection is not involved and they form only when the layers of air are stable. When there are several humid layers, one above the other, it is quite common for lenticular clouds to appear vertically 'stacked' one above the other: when they are closely spaced in a stack, they are known as a *pile d'assietes*, French for 'pile of plates'.

Polar stratospheric nacreous clouds

High-altitude wave clouds, known as polar stratospheric clouds (which form at altitudes of 15-30 km/16-18 miles) resemble cirrocumulus lenticular but are very beautiful pastel-coloured clouds known as nacreous, or 'mother-of-pearl', clouds. They are created when there is wave motion at their altitude, caused by the flow of air over mountain peaks below. Nacreous clouds are sometimes visible at high latitudes, most often 50°N or S, just after sunset and before sunrise when the sky itself is fairly dark, but when, because of their altitude, the nacreous clouds are illuminated by the Sun. The colours arise through defraction – waves of light being 'deviated' from their paths by cloud particles. The wavelength of light that is being defracted, and therefore, its apparent colour, depends on the size of the cloud particles. In nacreous clouds the colours are especially strong because the different parts of the cloud contain large numbers of particles all roughly the same size, which produces the bands of colour. Although beautiful, nacreous clouds are a cause of concern for meteorologists: the cloud particles are the 'laboratory' for the chemical reactions that destroy ozone. In the past, displays of nacreous clouds over the polar regions were seen only every 30 to 40 years, but in 1999 several displays were reported in that year alone – when the 'hole' in the ozone layer over the North Pole first became apparent.

BILLOWS

Technically known as the undulatus variety, billows are a quite common, though no less interesting and attractive, cloud structure. Billows generally arise when there is wind shear between two adjacent layers in the atmosphere – in other words, when there is a difference in wind speed or wind direction, or both, in the two layers. A layer of cloud is broken up into regularly spaced 'rolls' of cloud, with the billows running at right angles to the wind. Sometimes, though, there are finer undulations which run parallel to the

ABOVE: *Billows arise when there is a difference in either wind speed and direction (or both) in two adjacent levels of the atmosphere.*

wind: these are known as corrugations. Billows do bear a resemblance to wave clouds, but there is a very important distinction: billows move downwind with the cloud layer, while wave clouds remain stationary in the sky.

CLOUD STREETS

A cloud street is formed when cumulus clouds form a line that runs downwind. Cloud streets are created when there is a constant wind speed and a constant source of warm thermals: single cloud streets are commonly seen trailing downwind of oceanic islands because the island, heated by the Sun, is constantly causing convection, which produces cumulus clouds – which are then blown downwind, and the whole process begins again.

It is also possible to have numerous streets where wide areas of sky are covered by whole parallel streets of clouds with clear, blue sky between them: the 'clear lanes' in the sky are generally twice the width of the lines of clouds. The clouds are a recognised variety of cumulus known as cumulus radiatus (Cu ra) and most often occur when the lowermost layer of air is unstable but is capped by a stable layer – what is known as an inversion. Convection takes place below the inversion, with warm air rising below the clouds and sinking more gently into the clear air between the streets. For the streets to persist, the convection shouldn't be too strong – which is why cloud streets are most often seen mid-morning – and the clouds should evaporate quite quickly. If convection becomes too strong, the thermals will break through the inversion making the cloud growth irregular, and the parallel pattern starts to break down. If the clouds don't evaporate, but instead start to persist, they will start to spread out underneath the inversion and give rise to a layer of stratocumulus (Sc).

FALLSTREAK HOLES

Sometimes when there is a very thin layer of altocumulus (Ac) cloud

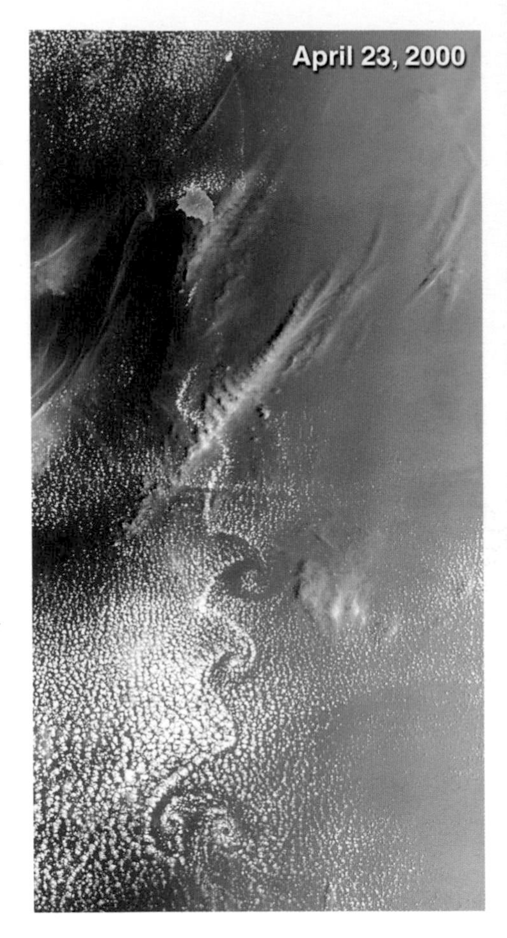

ABOVE AND RIGHT: *These MISR nadir-camera images from April 23, 2000 and May 9, 2000 show cloud swirls, like delicate lace, forming patterns known as von Karman vortex streets. The turbulent atmospheric eddies form in the wake of an obstacle, in this instance the 1050 m (3440 ft) high summit on the island of Socorro, Mexico. The surrounding clouds make the vortex patterns visible. To the northeast, much subtler disturbances are associated with the tiny Isla San Benedicto. Both islands are part of a group known as the Revillagigedo Archipelago, and are located about 400 km (250 miles) equatorward of the southern tip of Baja California.*

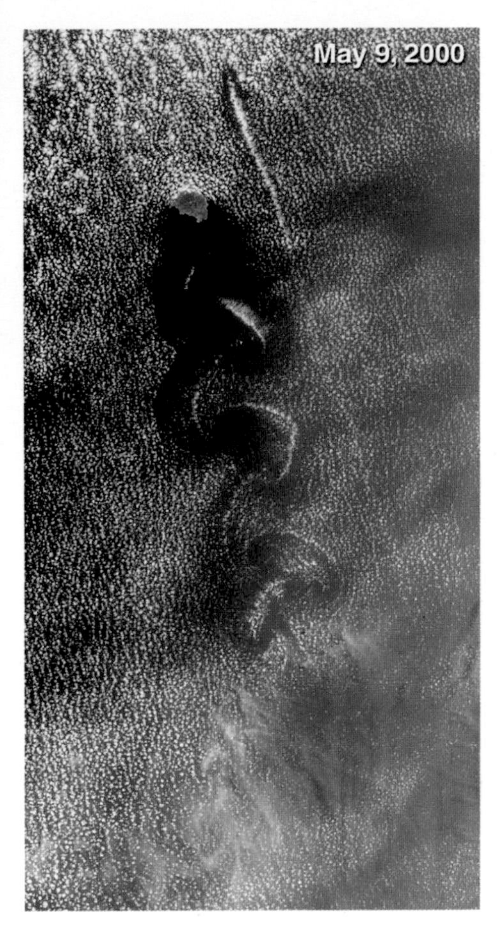

covering the sky, an isolated hole suddenly appears. These holes, known as fallstreak holes, are often symmetrical, appearing as circles or ellipses. Altocumulus clouds consist of super-cooled water droplets – in spite of being at temperatures of less than 0°C (32°F), the cloud droplets aren't frozen. Suddenly, freezing begins in part of the cloud layer and spreads, and the result is that the droplets, which are now ice crystals, fall to a lower level. Sometimes the ice crystals melt and evaporate as they reach lower, warmer levels, which means they become an invisible vapour, with no sign of their presence except for the hole itself.

CONTRAILS AND DISTRAILS

It's not certain what causes the sudden freezing that sets off the chain of events leading to a fallstreak hole, but a similar phenomena can be seen when an aircraft causes a thin layer of cloud to disperse in what is known as a dissipation trail, or distrail.

Distrails are less common than contrails – the condensation trails that are a very familiar sight in the sky. Contrails are the 'man-made' clouds caused by the condensation of water vapour emitted by aircraft engines. Water vapour itself is a invisible gas: when it condenses into cloud we see it in its new form. That's why there's always a clear space immediately behind the plane before the contrail 'cloud' begins.

The less common distrails occur when an aeroplane flies at the same level, or just above, a thin layer of cloud and 'slices' through it creating a clear lane through it. This can occur in one of three ways. First, the aircraft can mix warm air from above the cloud down into the cloud layer, causing the cloud to evaporate. Secondly, the heat from the engines may be enough to evaporate the cloud droplets. Thirdly, the original cloud may consist of super-cooled droplets, and solid particles emitted by the engine may act as a catalyst causing the super-cooled droplets to freeze and fall as ice crystals, which as they warm up, condense and turn to vapour, thereby leaving a clear lane – a bit like a fallstreak hole, but in a 'lane' following the plane's direction across the sky.

PYROCUMULUS

Cumulus clouds that are created by convection happen when there is a source of heat and water. Pyrocumulus, a name derived from the Greek for 'fire' and the Latin for 'heap', can often be seen above the cooling towers of power stations, but are also clouds that are created by fires. The fires don't need to be big: when farmers burn stubble after the harvest the heat may be enough to form small cumulus clouds at the top of the column of

MAIN IMAGE: *A contrail – condensation trail – is a man-made cloud caused by water vapour emitted by an aircraft engine.*

smoke. But large fires, such as those caused by natural wildfires or by burning forests to clear the land, can create extremely large cumulus clouds. Some cumulus become so large that they reach high into the atmosphere where they turn into cumulonimbus – the clouds that produce heavy rainfall but that also turn to thunderstorms. If the conditions on the ground are exceptionally dry, lightning from cumulonimbus produced over a wildfire has been known to strike and create new wildfires a considerable distance away from the 'parent' fire.

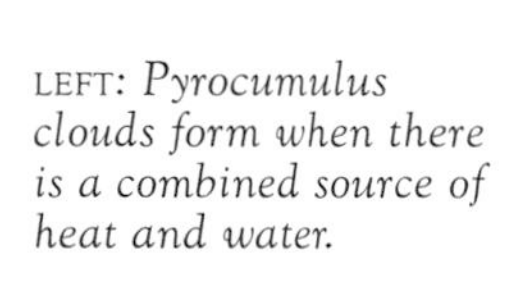

LEFT: *Pyrocumulus clouds form when there is a combined source of heat and water.*

CLOUDS AT NIGHT: NOCTILUCENT CLOUDS

Unless they are illuminated by the Moon, the vast majority of clouds are invisible at night. However, one type of cloud can be seen around midnight from certain parts of the Earth, around the time of the summer solstice. These are noctilucent clouds (NLC): their name means 'shining at night' and they are the highest clouds in the atmosphere, occurring just below the mesopause (the upper boundary of the mesosphere) at altitudes of around 80-85 km (50-53 miles). In high latitudes, beyond 45°N or S, for about one month before and after midsummer, the clouds are illuminated by the Sun, while the observers on the surface of the Earth are in darkness! Around midnight, noctilucent clouds have a characteristic bluish silvery-white colour; earlier in the evening, or later towards dawn, they may appear slightly tinged with yellow because the sunlight has travelled through a greater length of atmosphere.

There are four main forms of

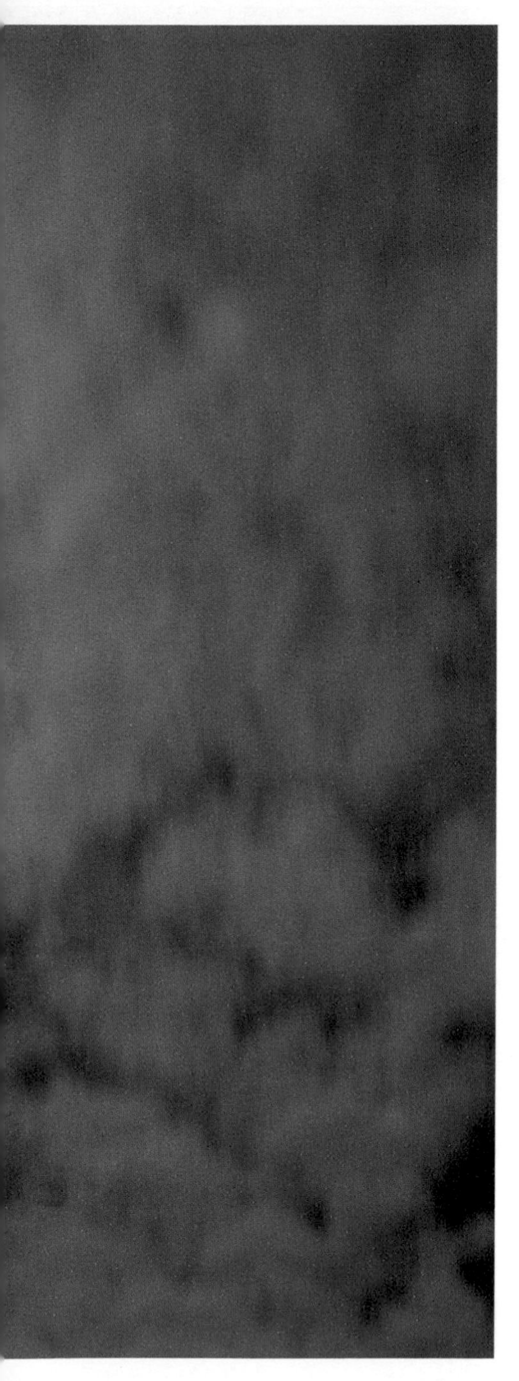

NCL: veils, which are described as 'tenuous films' and resemble cirrus (Ci) or cirrostratus (Cs); bands, which are long parallel streaks that sometimes change in brightness over periods of 20-60 minutes; billows, which are closely spaced, parallel, short streaks which can appear on their own or cross over long bands, and alter in brightness within minutes; and whirls, which are partial or complete rings or loops of clouds with a dark centre.

CLOUDS AND RAIN

All clouds are made up of tiny droplets of water: they might be liquid, super-cooled, or frozen, but they are so small that at least one million of them are needed to make a single raindrop. Most, but not all, precipitation forms in clouds: not all of the 'wet' that we experience in the weather is derived from clouds. Meteorologists define precipitation properly as water either in liquid or solid form that is derived from the atmosphere and falls through it, landing on the surface of the Earth. But, as we shall see, there is also water in liquid or solid forms that forms directly on the ground.

LEFT: *Usually clouds are only visible at night because of moonlight.*

PRECIPITATION

PRECIPITATION IS ANY FORM OF WATER, or ice, that is deposited on the ground – or any other surface. It can be divided into two categories: particles that fall through the atmosphere, and deposits that form directly on the ground. The particles that fall through the atmosphere include rain, drizzle, snow, sleet, and hail, while precipitation that forms on objects on the ground includes dew, hoar frost, rime and glaze. Although mist and fog are not technically precipitation, they are dealt with here because they are caused by suspended water droplets and occur, like cloud, when the air is cooled to dew point: in fact, they could be considered as 'clouds at ground level'.

Condensation, Stable and Unstable Air

CLOUDS FORM WHEN WATER vapour in rising air condenses into tiny water droplets or freezes directly into ice crystals. The height at which this occurs depends on the stability of the air and the amount of moisture present. Air is warmed when it is in contact with the ground, which causes it to expand and become less dense than the cooler air above it, and so it starts to rise. As the air moves upwards, its temperature begins to drop at a set rate, the lapse rate, making it denser and heavier as its height increases. Eventually, the rising air reaches a point at which it is more dense than the surrounding air, and it begins to sink back down again. This air is said to be stable. As air rises and cools, the amount of water vapour it can hold decreases, until, like a bath sponge, it has 'soaked up' as much water as it can hold and has become saturated. At this 'saturation point', water droplets form: the temperature at which this happens is called the dew point. The height of the dew point is not fixed but varies according to temperature. When water vapour condenses into water droplets it releases its latent heat: if the saturated air is cooling at a slower rate than the air surrounding it, it will also be warmer and continue to rise.

CLOUD CONDENSATION NUCLEUS

In order for precipitation to occur, there must be a way for the cloud droplets or ice crystals to grow large enough and heavy enough for them to fall as rain, drizzle, snow or hail. We all know that in 'normal' conditions, water freezes at 0°C (32°F): in the atmosphere, however, where water particles exist as minute cloud droplets, this is not often the case! At very high altitudes in the troposphere – around 10 km (6 miles) above the Earth's surface – many of the cloud particles remain in a liquid, 'super-cooled' state. Except at extremely low temperatures – around -40°C (-40°F) – liquid water will not freeze unless minute solid impurities are present. These minute nuclei come from many sources such as volcanic ash and soot, smoke, sea salt, and sand.

RIGHT: *Clouds*

The amount of these nuclei varies from ocean to continent, and in height within the troposphere: at sea level there are around 100-200 million nuclei in every cubic metre of air!

All cloud droplets have a nucleus around which they have condensed: the cloud condensation nucleus, or CCN. A typical CCN has a diameter of around 0.2 microns: a typical cloud droplet measures about 20 microns in diameter, and a typical raindrop measures a relatively enormous 2,000 microns in diameter!

Making raindrops

Cloud droplets are so small that growth by condensation is very slow, so other processes are needed to create raindrops. The two main processes by which a cloud droplet grows into a raindrop are freezing and coalescence (or collision), and meteorologists often speak of them as giving rise to 'cold rain' and 'warm rain' respectively.

COLD RAIN: THE BERGERON PROCESS

The 'freezing' process of producing precipitation is known as the Bergeron Process (and 'Ice Crystal Theory') and was first proposed by the Swedish meteorologist Tor Bergeron in 1933.

Like condensation, freezing most often needs the presence of a suitable nucleus. While the impurities that are found in air are very numerous at sea level, higher up in the troposphere they are less numerous. In the atmosphere, only one cloud droplet in a million is frozen at -10°C (14°F), and a couple of hundred or so in every million are frozen at -20°C (-4°F).

ABOVE: *A typical drop measures about 2 microns in diameter.*

At -40°C (-40°F) and below all the cloud droplets are ice crystals – the temperature at which liquid water freezes without needing a nucleus around which to form. Ice crystals that form from vapour alone take on characteristic shapes depending on the temperature range within which they are created. As the crystals fall through progressively warmer layers, they become more complex in shape (they can also change shape if they are carried upwards on up draughts into cooler regions of clouds).

Depending on the temperature range, the ice crystals may be flat hexagonal plates, hollow hexagonal columns, or dendrite crystals of varying complexity. Dendrite means 'branching like a tree' so dendrite crystals are probably best described as a flat hexagonal plate with six 'branches' growing out from the centre.

The best surface on which to freeze water is an ice crystal, so if super-cooled water droplets touch one of these, they freeze instantly. This means that in mixed clouds such as altostratus (As), nimbostratus (Ns) and cumulonimbus (Cb) where there are both super-cooled water droplets and ice crystals, the crystals will grow rapidly. This freezing of super-cooled water on to ice crystals is known as riming (and is basically the same process that causes the deposit of rime as frost on the ground). Ice crystals formed in this way are known as graupel. In the atmosphere, the ice crystals grow at varying rates depending on just how much super-cooled water lands on them. Larger ice crystals can also 'capture' others, because bigger crystals fall at higher speeds than smaller ones. The size of the crystals, which melt when they fall into warmer air, and the temperature of the air through which they are falling, determines the form of the resulting precipitation – rain, snow or hail – which reaches the surface: tiny ice crystals that have melted into smaller droplets often evaporate before they reach the

ground, but larger droplets survive to fall as cold rain. In winter months, quite shallow cumulus clouds can become glaciated and give rise to a cold shower. In summer, though, immense towering cumulus congestus may not turn into cumulonimbus because even their high-altitude tops don't reach a sufficiently low temperature.

WARM RAIN: COALESCENCE

In contrast to the 'cold clouds' that produce ice crystals, other clouds whose tops are warmer than -15°C (5°F) often generate precipitation in different ways. Coalescence is the term used by meteorologists to describe the process by which droplets collide and grow.

The 'warm' clouds are made up of cloud droplets of varying sizes: a typical cloud droplet measures a mere 20 microns in diameter and has a fall rate of 0.01 m (1/2 in) per second; a large cloud droplet can measure 100 microns in diameter and has a fall rate of 0.27 m (10 1/2 in) per second. The larger drops fall faster than the smaller drops and so tend to overtake and collide with them, assimilating them and therefore growing: a small drizzle drop measures some 200 microns and has a fall rate of 0.70 m 27 1/2 in) per second; a typical raindrop measures 2 mm (1/8 in)in diameter and has a fall rate of up to 6.5 m (21 ft) per second.

LEFT: *Scientists believe it takes between 20 and 60 minutes for an average-sized raindrop to form. Average is 1-2 mm (about 1/8 in) in diameter, although large drops that are distorted by atmospheric pressure have spherical 'equivalent' diameters of 5-8mm (about 1/4 in).*

The number of raindrops formed within the clouds depends on the liquid water content, the range of droplet sizes, and the strength of the up-draught inside the cloud as this determines the amount of time that is available for a droplet to grow. Even if the droplets do coalesce to become raindrops and fall out of the clouds, there is some evaporation taking place in the layer between the cloud base and the surface, the sub-base. If the air in this layer is dry and the raindrops small, they may completely evaporate on their way down to the ground.

The process of coalescence is most likely to take place in very deep clouds, such as cumulus congestus found in summer months: scientists have calculated that it takes between 20 and 60 minutes for an average-sized raindrop to form, so shallow cumulus clouds or stratiform clouds such as stratus and stratocumulus do not produce large raindrops or great

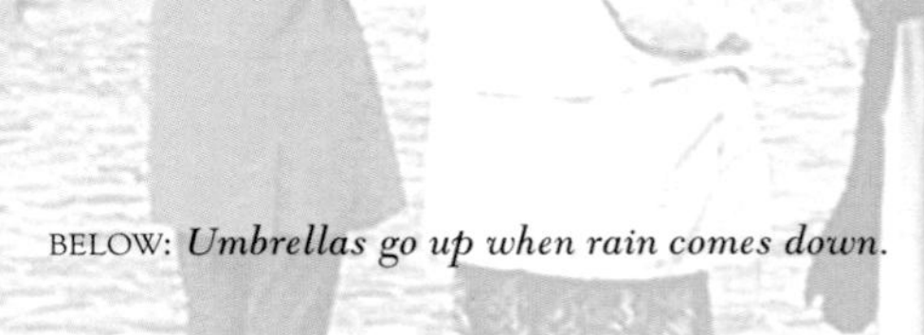

BELOW: *Umbrellas go up when rain comes down.*

quantities of rain, but they may give rise to fine drizzle. Drizzle is rainfall whose droplets are less than 0.5 mm (1/3 in) in diameter. When stratiform clouds do appear to have brought rain, it's more likely that they have been 'seeded' by ice crystals that have fallen from higher cloud layers such as altostratus.

ABOVE: *Approaching storms can give rise to heavy bursts of intense rain or hail.*

SHOWERS AND BANDS OF HEAVY RAIN

Rain from cumulus congestus and cumulonimbus occurs in the form of showers, which, while they may be heavy, are generally restricted in duration and in location: rarely does this type of rain shower affect an area of land more than a few kilometres wide. With clusters of cumulonimbus, there may be several cells producing rainfall, and so a larger area may be affected.

In total contrast to these showers is the rain produced at frontal systems. Here, the main rain-bearing clouds are altostratus, nimbostratus and, to a lesser extent, thick stratocumulus. An area affected by frontal rain may be several thousand square kilometres and this often means that the heavy rain tends to be arranged in bands running roughly parallel to the front, with areas of lesser or no rainfall between the bands. In a frontal system, it may be possible for cumuliform clouds to be embedded in the stratiform clouds, which may give rise to heavy bursts of intense rain or hail.

Hail

HAIL IS MADE UP OF large pieces of ice that form inside and fall from cumulonimbus cloud. A hailstorm can often occur on a summer afternoon and can be devastating. In northern France in July 1788, hailstorms destroyed the wheat crops, causing widespread shortages. Exacerbated by the severe winter that followed, shortages turned to famine which led to social unrest. Consequently, many historians consider the effects of the weather at this time to be a major contributor to the French Revolution. In the USA, hail causes about $100 million worth of damage to property and crops each year: the most expensive hailstorm in the USA to date occurred in Denver, Colorado, on 11 July 1990, when the total damage cost an estimated $625 million.

GROWING HAILSTONES

Hailstones consist of alternating layers of clear and opaque ice. They are formed from graupel, 'soft hail', which consists of white opaque ice particles. Graupel, which is often snow as 'snow pellets', is nearly always spherical, but it can be conical in shape, too, with a snow-like structure and a diameter of up to 5 mm (1/3 in). Graupel acts as the nucleus, and small hailstones grow due to the accumulation of super-cooled water droplets as they are carried upwards on rapidly ascending air.

The layers are in fact 'grown' at the different regions and different temperatures within the cloud: the opaque layers are from the super-cooled water which freezes on impact and then traps a layer of air between them. The clear layers are formed in the warmer regions of clouds when liquid droplets coalesce over the surface of the hailstone before freezing to give what is, in effect, a layer of glaze.

The conditions for hailstones to grow are found deep inside the cumulonimbus clouds where the strong up-draughts are 'tilted', allowing the growing hailstones to be carried upwards, thrown out of the up-draught at high level, and then to fall back to the surface – only to meet another up-draught at a lower level and repeat the process. A hailstone can go 'up and down'

RIGHT: *Technically, hailstones have a diameter of 5 mm (1/3 in) and they can 'grow' as large as 14 cm (6 in).*

through a cumulonimbus cloud several times in this way: counting the layers of ice in a large hailstone is the way to find out how many times it has made an ascent! Eventually, the hailstones become too big to be supported by the up-draught, and they fall to the ground. The first sign on the ground that hail might be on its way is a growing whitening among the shafts of rain. The next sign is audible: hailstones striking roofs, pavements and car windscreens!

SMALL HAIL AND HAILSTONES

Weather lore is full of accounts of incredibly large hailstones varying in size from oranges to elephants! Like drizzle and rain, meteorologists distinguish between 'small hail' and 'hailstones'. Small hail is about the same size as graupel – up to 5 mm (1/3 in) in diameter – but has a higher density. These particles are generally semi-transparent and rounded with conical tips, and often consist of liquid water contained in a frozen outer 'shell'.

Hailstones have diameters greater than 5 mm (1/3 in). They can be the size of peas or as large as grapefruits, 12 cm (5 in) in diameter! In Britain, the largest hailstone, which fell in 1980, weighed in at 626 g (22 oz) with a diameter of 9.5 cm (3.75 in). The largest hailstone to fall in the USA dropped on Coffeyville, Kansas in September 1970: it weighed 757 g (27 oz) and had a diameter of 14 cm (6 in).

Individual hailstones up to 1 kg (2 lb) have been recorded: these big hailstones are created by super cell storms which contain the necessary and exceptionally strong up-draughts to carry larger hailstones in their ascent. Super cell storms may produce devastating hailstorms which cause extensive damage and prove fatal to animals and humans. Even bigger than individual hailstones are hailstone aggregates – where several hailstones have joined together and become frozen into a single mass. Most often, the sheer size of aggregates causes them to smash on impact with the ground. China and Bangladesh regularly experience aggregates of hailstones, but so far, the record goes to an aggregate which fell on Kazakhstan in 1959 and weighed an incredible 1.9 kg (4 lb)!

MIND YOUR HEAD!

In North America, hailstorms reach their peak across the high plains stretching from Texas to Alberta between May and June, although April and July can be busy months too! On average, May and June will experience more than 2,000 storms producing hail the size of golf balls, with diameters of more than 50 mm (2 in)!

Hailstones are fast-moving: they fall from clouds at speeds between 30 and 90 km/h (20-55 mph). A single, solid lump of ice about the size of a golf ball travelling at 90 km/h (55 mph) from a great height may do some serious damage if it happens to hit you! Indeed, hailstorms have been known to kill: the South Carolina Gazette reported in May 1874 that 8 people had been killed in a hailstorm along the River Wateree; in 1888 in northern India, 246 people (and 1,600 sheep and goats) were reportedly killed by hailstones the size of cricket balls; and in 1932, in western Hunan Province in China, 200 people were killed and thousands injured in a severe hailstorm.

Snow

A SOFT BLANKET OF SNOW across the landscape is both a blessing and a curse: snow creates not only an incredibly beautiful winter wonderland scene, but as snow accumulates on mountains in winter, it forms the 'frozen reservoirs' that are vital for summer irrigation. But snow can also be one of the weather's most destructive forces, bringing disruption, chaos and, sometimes, death.

If the ice crystals that are formed high in the cloud melt on their way down to the surface, rain falls. If they don't melt, the original ice crystals with their hexagonal structures may collide with each other and become frozen together. This happens when the temperature in clouds is just below 0°C (32°F), when the thin film of water on the surface of the crystals freezes and locks them together. It's these aggregates of many ice crystals that form large 'wet' snowflakes which float to the ground.

LEFT: *Skiiers love dry,powdery snow.*

Wet snow melts under the slightest pressure, but it refreezes as soon as the pressure is removed, and becomes a compacted sheet of ice. It's this aspect of wet snow that causes headaches for transport providers and hazardous conditions for drivers (and pedestrians). Wet snow is the type of snow generally experienced in Great Britain: because it can't be removed by being 'blown away' it's much harder, and much slower, to clear away 'by hand' as wet snow weighs about 350 kg for every cubic metre! That's why, in Britain, if snow falls unexpectedly overnight, and the roads and railway lines haven't been cleared and treated with salt and grit to break up the compacted ice sheet formed under pressure, most people arrive late at work – and just in time to start the struggle homeward!

At very low temperatures, the ice

ABOVE: *Black ice is formed when wet snow melts under slight pressure then refreeezes, once the pressure is removed, to become a compacted sheet of ice.*

crystals remain separate and fall to the ground as powder snow – dry, powdery snow which is not 'packed down' – much to the happiness of winter sports enthusiasts (and transport providers!) Powder snow falls at sub-zero temperatures; it is dryer and less dense than wet snow, weighing about 110 kg per cubic metre, and can be 'shifted' by using snow-blowers.

In order for snow to fall there has to be a constant inflow of moisture to feed the growing ice crystals: an air stream might pick up moisture as it passes over a relatively warm water surface such as a large lake or ocean. If the moisture is lifted to higher and cooler regions of the

NEXT PAGE: *A blanket of snow can be both beautiful and destructive.*

LEFT: *Snow blower at work: this kind of equipment can only be used to remove dry snow. Wet snow is much heavier, and cannot be blown away.*

atmosphere, then snow may occur. Mountain areas can also initiate snowfalls by orthographic lifting of a moist air stream in below-freezing temperatures.

FLURRIES, SLEET AND BLIZZARDS

Snowfalls come in a range of 'shapes and sizes' from snow 'flurries' – light, brief 'snow showers' – to heavy falls of snow. Snow flurries have their own hazards as they may reduce visibility to less than 200 m (650 ft). In Britain, the accepted term for partially melted snowflakes is sleet – a mix of snow and rain. In North America though, sleet is also the term used to describe wet snow and ice pellets – frozen raindrops, or melted and refrozen snowflakes. In both cases, the particles either began their life in, or passed through, a warm layer of air before falling into a colder layer where they froze.

ABOVE: *Blizzards can cause chaos, and can be dangerous.*

What makes for 'heavy snow' varies from region to region. In places where snowfalls of 10 cm (4 in) are common, heavy snow would be around 15 cm (6 in). In cities with large built-up areas and heavy traffic keeping the temperature higher, heavy snow may be a mere 5 cm (2 in). In both cases, heavy snow means it's going to disrupt normal routine! The world record for annual snow total belongs to Paradise Ranger Station on Mount Rainier, Washington State, USA. Between 19 February 1991 and 18 February 1992, Paradise recorded a total of 31.12 m (over 60 ft) of snow!

Blowing and drifting snow usually happen when strong winds move falling snow or loose snow on the ground. Blowing snow is snow lifted from the surface and blown about, significantly reducing visibility. Drifting snow is when strong winds are blowing loose snow into piles or drifts. The most perilous of winter storms though are blizzards, when low temperatures combine with strong winds (Force 7 on the Beaufort Scale) bearing large amounts of fine powdery

BELOW: *In Britain, the accepted term for partially melted snow is sleet, while in the US, this term describes wet snow and ice pellets.*

ABOVE: *Snow piled in the streets after New York City blizzard, January 2005.*

particles that reduce visibility to only a few metres.

AVALANCHE!

Freshly fallen snow contains a significant amount of air but, after a while, the weight of the snow becomes compacted and begins to change its nature: larger crystals grow at the expense of smaller ones and this can create a weaker layer lying just below the surface. This is because the temperature varies at different levels of snow cover: if the snow is thick enough, the heat and energy stored up in the ground over the summer will keep the base of the snow at 0°C (32°F). Insulated by the blanket of snow, the temperature at the base of the snow is warmer than that at the top. In mountainous areas and in northern latitudes, many animals can survive by digging a deep hole into the snow to keep warm at night when surface temperatures can be 10-30°C (20°-80°F) lower than deep under the snow!

Inside the blanket of snow, where there is a slightly higher temperature, crystals begin to evaporate and produce water vapour. As the face of each 'warm' crystal evaporates, water vapour is released which then freezes on to the colder face of the neighbouring crystal. This process is known as the vertical temperature gradient. If the vertical temperature gradient is high, large, cup-shaped crystals will form as a result of the process of transferring water vapour and frost. A slippery layer of snow can then form under the surface of the snow blanket and increase the risk of avalanches, when the uppermost layers of snow become detached and slide over the weaker layers beneath. Avalanches can also occur when a thaw sets in, either due to a rise in daytime temperatures, Sun partially melting the snow, or heavy rainfall. When the snow quickly refreezes at night, the structure of the snow is altered. Avalanches can also be 'triggered' by vibrations – natural or human: even an off-piste skier can help set off an avalanche in unstable snow.

RIGHT: *Avalanches can result from changes in the verticl temperature gradient, from thaws, or can be triggered by vibration.*

Dew

ABOVE: *Dew, unlike rain, does not have origins in clouds.*

UNDER CLEAR SKIES, objects rapidly lose their heat after the Sun goes down because they radiate their heat back into space. When surfaces on or near the ground – such as blades of grass, or leaves – cool to below the dew point (but not below freezing point) the invisible water vapour in the air condenses on them, depositing tiny droplets of visible water, or dewdrops, on the surface. Dewdrops are relatively small – below 1 mm (0.004th in) in diameter. Unlike rain, dew does not have its origins in clouds: the source of the water vapour is more often the underlying soil, which is partly insulated by the air lying between the leaves and the ground, and therefore remains slightly warmer than higher surfaces. In regions that are arid by day, and where precipitation is very sparse, morning dew is a vital source of water for plants and animals.

Not all water droplets on plants are dewdrops, however. When the air is very humid, guttation drops often form at the tips of leaves and blades of grass. The drops are larger than dewdrops – about 2 mm (1/8th in) in diameter – and in them can often be seen a tiny 'rainbow'. They are the result of the plant's transpiration, the process by which a plant moves water from its roots to its leaves, where the water is then evaporated. In humid conditions, the water transportation to the leaves continues, but the water vapour cannot evaporate into the air because the air is already saturated.

HOAR FROST

If dew has formed and the temperature drops, dewdrops don't necessarily freeze immediately, but exist in a super-cooled state below 0°C (32°F). This super-cooled state can often be seen when fern-like patterns of frost form on the insides of windows and on the exteriors of car windscreens. Once a few ice crystals have formed, they start to 'grow' across the surface, replacing the super-cooled droplets, which evaporate. Super-cooled droplets will freeze if

RIGHT: *Frost on berries: if dew has formed and the temperature drops, the dew can freeze. This is known as 'white dew' or 'silver frost'.*

temperatures drop to around -3° to -5°C (23° to 26°F). Liquid dewdrops that have frozen because of a drop in temperature are known as white dew or silver frost.

The most common form of frost is hoar frost. Hoar is the equivalent of dew but importantly, the water vapour has been deposited as ice crystals – it has never been through the liquid stage. Like dew, hoar frost develops under clear, calm conditions. The temperature to which the air must cool to produce frost is not the dew point, but the frost point: the air must be cooled at a constant pressure and humidity in order to reach saturation with respect to an ice surface, rather than a liquid water surface.

BELOW: *Sharp needles of rime are deposited by super-cooled fog.*

RIME

Looking superficially like hoar frost, rime is a less common form of frost. Rime is a white deposit on exposed objects, and it is formed by a completely different method to hoar frost. Rime is deposited from super-cooled fog and the tiny water droplets remain liquid until they come into contact with an obstacle, when they freeze instantly. Most often, long 'feathers' of rough ice build up on the windward surfaces of objects (the 'feathers' pointing into the wind). Although fog is generally associated with fairly light winds, sometimes if the fog hangs around for long enough, large amounts of ice can be deposited, especially in mountainous regions that are shrouded in super-cooled air for long periods. Under still, near

windless conditions, needle-like crystals often grow around the edges of leaves and similarly sharp-angled forms. In mountainous regions, the rime accumulations may be several metres across, and these can break away and fall into the valleys below, where they add to the ice that is found in valley glaciers. Large accumulations of rime are also often found on very tall radio and television masts, and large deposits on trees can cause physical damage, but this is never really as severe as the damage caused by deposits of freezing rain known as glaze.

ABOVE: *Ice storms are created by slow-moving frosts and are very destructive.*

GLAZE

Glaze is commonly known to many people – motorists in particular – as 'black ice'. When drizzle, raindrops or fog freezes on contact with a surface, a transparent layer of ice, called glaze, can be formed. Because it is transparent, and therefore not especially visible, glaze becomes a hazard. To drivers, the road appears to be wet, rather than frozen. Although the water droplets freeze very rapidly when they land on a surface, they still have just enough time to spread themselves out into a thin layer before they freeze. Glaze most often occurs ahead of an approaching warm front in a depression: when the temperature of the air ahead of the front is below freezing, any rain that falls from the warm moist air overhead is transformed into glaze when it hits the ground – or any other surface. When a front is slow-moving, or stationary, for a long period, layers of glaze can form and give rise to what are called 'ice storms'. Ice storms are generally sudden, widespread and unfortunately, highly destructive events. For example, between 5-9 January 1998, an ice storm hit large parts of Canada and New England, and the accumulated weight of glaze broke down trees, telephone and power lines and caused electricity pylons to collapse.

Visibility: Haze, Mist and Fog

NOT TECHNICALLY precipitation, mist and fog are what meteorologists call 'obscurations' of the air – they 'obscure' or reduce visibility. Haze is also an obscuration, but this is caused by dry articles that are small enough to remain suspended in the air for long periods. Haze may give a rather pearly appearance to daylight, but there are no specific upper or lower limits to visibility, as haze rarely reduces visibility below 1 km (2 miles). The official definition of fog is a condition where the horizontal visibility is 1,000m (3330 ft) or less because of the presence of water droplets suspended in the atmosphere. Thick fog is when visibility is 100 m (330 ft) or less; impaired visibility of more than 1,000 m (3330 ft) is mist.

The process by which mist and fog are produced is identical to the formation of clouds, except that it occurs at ground level. The air in the atmosphere is composed primarily of a combination of dry air and water vapour that is present in an invisible gaseous form. Sometimes, when the air becomes saturated with water vapour, some of the water condenses into thousands of microscopic liquid droplets and remains suspended in the air. If

MAIN IMAGE: *An 'obscuration of the air' is what meteorologists call fog and haze.*

there are enough of these droplets, they will become visible to the eye in the form of a cloud – or fog. The two most common types of fog are radiation fog and advection fog, but there are also other, rarer, types such as steam fog, known as Arctic Sea Smoke, and ice fog.

RADIATION FOG

Radiation fog is created when the Earth's surface radiates heat into space, cooling the air above it. In order for the heat to radiate away, the sky needs to be clear of clouds, there needs to be a long-enough period for temperatures to fall, and the air needs to be moist – which is why long, clear autumn and winter nights after a day of rain are the best ones for producing radiation fog! The same moist conditions may be found in river valleys and near large bodies of water. Radiation fog does not form over seas and oceans because these bodies of water don't cool rapidly after sunset like the land does. The final ingredient needed to create radiation fog is a light wind, no more than 4 knots (7.5 km/h/4 mph) which is about 'walking speed'! This light wind ensures that the cooling process is spread gently throughout the lowermost layer of air: any stronger wind would blow the fog away!

Britain has a largely undeserved reputation for being foggy: Shakespeare wrote of 'seasons of mists and mellow fruitfulness' but it was probably Charles Dickens who firmly established the idea of London forever enveloped in thick 'pea souper' of fog! In fact, compared with Continental Europe, Britain has a fairly ordinary

BELOW: *Radiation fog is created when the Earth's surface radiates its accumulated heat into the air above it.*

experience of fog – only about 40 days per year. It generally occurs in a broad belt running from Newcastle in the northeast, down through the Midlands and into the Sussex Weald in the southeast of the country. The beautiful Pembrokeshire coast of Wales and the west coast of England – especially around Land's End in Cornwall, are more frequently affected by sea fog.

'A foggy day in London town' when, as the song goes, 'The British Museum has lost its charm', is a pretty rare event these days! Belgium, the Netherlands and parts of northeastern Germany experience about the same number of foggy days as Britain each year, but France is by far the foggiest place in Europe! Mont-de-Marsan in the Landes region has fog with visibility of less than 1 km (0.6 mile) on average 105 days per year; Rouen often has its lovely cathedral shrouded in fog for some 97 days a year (which probably explains why the Impressionist painter Claude Monet was so keen to capture the effects of sunlight on its facade at different times of day!), while Lille and Rennes can expect between 70 and 80 days of fog annually.

When conditions are right, fog can form very quickly and visibility can drop from 3-4 km (2-3 miles) to under 200 m (660 ft) in less than 10 minutes. Radiation fog is dispersed during 'insolation' – warming by the

ABOVE: *Fog in mountains.*

Sun. As dawn breaks, the Sun begins to warm the ground, which, in turn, then warms the air above it. Alternatively, radiation fog may be dispersed if the wind rises and 'blows it away' by mixing the layer of fog with drier, overlying air. In both cases, the fog may lift into a layer of cloud, which may hang around all day!

Very shallow ground fog – about 1-2 m (3-6 ft) in depth – sometimes forms very quickly around sunset, and often after late afternoon rain. In this case, the underlying ground is not cooling, but the lowermost layer of air is. It is also possible for the water droplets in fog to be super-cooled and exist unfrozen at temperatures at below 0°C (32°F). When super-cooled fog drifts across the ground, any of the super-cooled droplets that come into contact with objects, such as leaves or grass, freeze on contact and form a deposit of rime.

ADVECTION FOG

Advection fog is produced when mild, humid air blows in from the sea and condenses over the cooler land along coastlines. All coastal regions are affected by advection fog, some more spectacularly than others: advection fog is a common

RIGHT: *Fog affects coastal regions. Mild humid air is blown in from the sea and condenses over the cooler land.*

but no less wonderful sight on the Grand Banks off Newfoundland, along the central and northern California coast and in the Bay of San Francisco, where it often tumbles in and obscures the Golden Gate Bridge. In the northeast of England and Scotland, advection fog is known locally as 'haar', while in Northumberland it is called 'fret'.

Unlike radiation fog, advection fog requires movement of the air: advection is a term used almost exclusively in meteorology and oceanography to refer to the horizontal movement that transports a substance (such as air) or some property (such as humidity) from one point to another. For example, 'thermal advection' means the amount of heat transported by the wind or by ocean currents. Consequently, advection fog moves. Even with strong winds of 30 knots (55.5 km/h/35 mph) over the seas, thick fog may still be present, but as the winds increase the fog may lift to form extensive stratiform cloud. Advection fog is found in areas of poleward moving tropical maritime air that is cooled by contact with the sea's surface (which is why it is also known as 'sea fog'), and it occurs most often in spring and early summer when the sea's surface temperature is at, or just recovering, from its lowest temperature. Extensive areas of sea fog can shroud exposed coasts. The fog may burn off inland when the ground warms up by insolation, but along the coast itself, the fog may persist all day.

ARCTIC SEA SMOKE AND ICE FOG

Meteorologists call this form of fog 'evaporation fog'. Any body of water, no matter how small, will have a direct influence on the air above it. The air temperature shifts until it is almost the same as that of the water and the air becomes charged with water vapour. If a wind then brings the air into contact with a cooler air mass, the water vapour will condense and form evaporation fog. Wreaths of 'steam' or steam fog, often about 50 cm (22 inches) high can form over a body of water, especially if the water is relatively warm, because water vapour rising from the surface condenses immediately. Steam fog can sometimes be seen over ice-free rivers and lakes on cold winter mornings, but it is at its most spectacular in polar regions where the sea is often warmer than the freezing Arctic air above. It is known as Arctic Sea Smoke.

Ice fog occurs in high latitudes in Siberia and North America as well as over Antarctica during calm, clear weather. It occurs when the temperatures drop so low – below -30°C (-22°F) – that the super-cooled fog droplets freeze into tiny ice crystals just 20-100 microns in

ABOVE: *Fog over ice occurs in high latitudes during calm weather, when temperatures drop below -30°C (-22°F). The super-cooled fog droplets freeze into tiny ice crystals, known as 'diamond dust'.*

diameter. (When the ice crystals are smaller than 10-20 microns in diameter they are called droxtals.) The crystals don't reduce visibility, but instead glitter like tiny diamonds in the sunlight, earning ice fog its popular name of 'diamond dust'. When sunlight is refracted through these crystals, brilliant optical effects are produced. It is to these, and many other, weather-related optical phenomena that we now turn.

OPTICAL PHENOMENA

A WIDE RANGE OF PHYSICAL CONDITIONS in our atmosphere are responsible for the myriad of forms and beautiful colours to be seen in different parts of the sky. Some of the optical phenomena, such as rainbows, different colours in the sky at dawn or dusk, and even mirages are common, but nonetheless beautiful. Others such as purple light and Bishop's Ring are much rarer occurrences, and in some instances, certain optical phenomena remain little understood.

SAFE VIEWING

IT CAN OFTEN BE DIFFICULT to see details of optical phenomena, or even clouds, especially when they are near the Sun, because the sky appears so bright. It may seem obvious, but it is vital that you never look directly at the Sun because eye damage could be the result. Instead, be safe, and 'block out' the Sun. You can use your hand or any other suitable object, but by far the best way to observe a sunlit sky is to view a reflection of it! A deep pool of still water that has a dark bottom is very effective, a piece of dark glass – even reflective 'mirror' sunglasses – all work well to view the Sun's reflection! In town and city centres, make use of the tinted glass on the exteriors of office buildings and watch the drama of the sky play out before your eyes!

ABOVE: ***Don't be tempted to look directly at the Sun, as it can cause permanent damage.***

Optical phenomena can be produced by light either from the Sun or the Moon, but those formed by moonlight are often fainter and can exhibit less, or even no, colour. In some instances, different optical phenomena may produce the same effects, so it sometimes can be difficult to distinguish between them. If a process of elimination is used, and there is also sufficient information about position in the sky and the angular size, it is often possible to 'work out' which phenomena is which. These 'elimination questions' include:

RIGHT: ***The safest way to view the sky is to look at a reflection.***

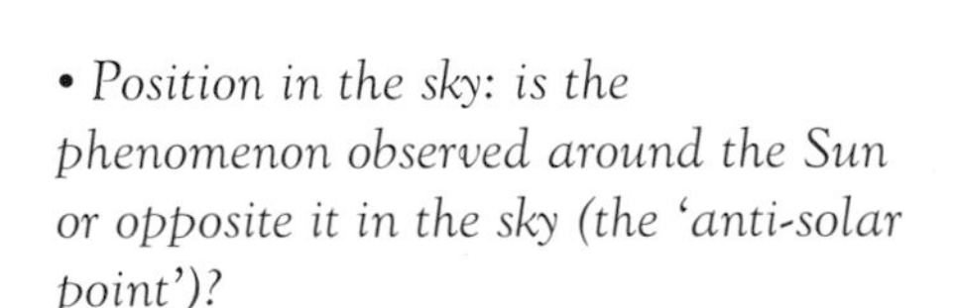

- *Position in the sky: is the phenomenon observed around the Sun or opposite it in the sky (the 'anti-solar point')?*

- *Colours: which, if any, are visible?*

- *If it is dispersed into spectral colours (the colours of the rainbow), does red*

or violet appear closest to the Sun?

- *If it is a spectrum (the colours of the rainbow), are the colours spread out horizontally, vertically, in an arc centred on the Sun, or in another more complex way?*

- *If a line or an arc, is it straight, a portion of a circle, an ellipse, or a more complex shape?*

- *Angular size: it's diameter, but also its angular distance from the Sun (see below).*

A very simple way of estimating the angular size of an optical phenomena is to use your hand: hold your arm out straight in front of you.

If the 'width' of the phenomenon you are observing is wider than one finger, the angular size can be roughly estimated at 1°; if the width is about the size of your clenched fist, then the angular size is roughly 10°; if the width is as wide as your hand with all your fingers stretched out like a star, then the angular size can be said to be roughly 22°.

FOUR MECHANISMS

ABOVE: *The sky reflected on the surface of still water is an example of specular reflection.*

MANY OF THE stunning visual phenomena result from the effects of solar beams passing through the tiny ice crystals or water droplets in the atmosphere. There are four basic mechanisms at work: reflection, refraction, diffraction and scattering – but some optical phenomena involve various combinations of the four mechanisms.

REFLECTION

Reflection takes place when a surface returns a portion of the incoming radiation back to its original source. A mirror, or the still surface of a pond, is an obvious example of how reflection works: your face is the 'source' and the surface of the mirror 'bounces' back your reflection. This type of reflection is known as specular reflection. In reality, reflection is a little more complicated. The exact amount of radiation returned to the source depends on the wavelength of the radiation, the nature of the reflecting surface, and the angle of incidence. In fact, the majority of natural surfaces give rise to what is called diffuse reflection, where the radiation is 'bounced back' towards the source in a wide range of directions.

ABOVE: *Rainbows occur when light is refracted from its original path.*

REFRACTION

Refraction refers to the deviation or 'bending' of a light wave from its original path when it passes from one medium to another. As light waves progress from their source to our eyes, the direction of the motion of the light waves changes. Refraction is a significant factor in the production of the halo phenomena including circumhorizontal and circumzenithal arcs, as well as parhelion, rainbows and dewbows. Refraction can also occur in the lowest layers of the atmosphere, where it produces mirages.

ABOVE: *A corona is a set of coloured rings around the Sun or Moon.*

DIFFRACTION

Diffraction is the dispersion or deviation of light waves as they meet the edge of a physical barrier, such as cloud droplets. The light rays are consequently 'bent' around the barrier. The most common effects of diffraction in meteorology are the optical phenomena such as Bishop's ring, corona, fogbow and glory, as well as producing iridescence and nacreous clouds (see page 54).

SCATTERING, OR WHY THE SKY IS (SOMETIMES) BLUE

This is the process by which small particles, such as molecules of gas, water vapour droplets, dust, smoke, or any other small particles that are suspended in the atmosphere, deflect the light waves that fall on them. Not all the wavelengths of light – and therefore the colours of the spectrum – are scattered equally: the wavelengths that are scattered depend on the size of the individual particles. The bulk of our atmosphere is composed of oxygen and nitrogen molecules: these scatter violet and blue light which have short wavelengths, but have little scattering effect on longer wavelengths (such as yellow, orange and red). The scattering occurs pretty much in all directions and so it is this scattered blue (and violet) light that makes the sky appear blue – humans aren't very sensitive to the violet light that is also scattered.

Larger particles, such as water vapour droplets, scatter longer wavelengths of light: the sky as seen from up a mountain or from the porthole on an airplane is cold and dry, and appears a deep blue colour. Descend to a lower, warmer and therefore atmospherically 'wetter' level, where the air is more humid, and the sky will appear a paler blue because the water vapour droplets have 'diluted' the pure blue.

BELOW: *A blue sky is the result of scattering.*

ABOVE: *Distant mountains appear a hazy blue due to atmospheric perspective.*

ATMOSPHERIC PERSPECTIVE

Except perhaps at the South Pole (where even the air close to the ground is cold and dry and so the sky appears a very intense blue), at low altitudes the air is rarely perfectly clean and dry. The sky is a paler blue and if we look at distant objects, such as hills or mountains, we see that they take on a pale-bluish tint and are less defined and distinct than objects closer to us. This effect, which has been captured on canvas by many great artists from Leonardo da Vinci (look at the background landscape in his Mona Lisa) to landscape painters such as John Constable and J.M.W. Turner, is called atmospheric perspective, and is the 'optical information' that our brain 'processes' in order for us to judge distances – to estimate how far away the hills or mountains are!

ONCE IN A BLUE MOON!

Dry haze particles, because of their size, also scatter long wavelengths and absorb some of the light. If the Sun is high in the sky, haze may give the light a 'pearly' quality. But if the Sun is low on the horizon, the haze often appears as a distinctive dark layer, often with a brownish tinge to it. A blue Moon – or a blue Sun – is not the fancy of songwriters and poets, but a rare phenomenon. Particles of soot and smoke, particularly from forest fires, fine dust from volcanoes and especially dust called loess which arises from the soil and is widespread over wide areas of Asia, are all of the requisite size (about 0.35-0.65 microns) to selectively remove all wavelengths of light except for blue and green. This process is known as preferential scattering and it makes the Moon or Sun appear blue and, very occasionally, green.

SUNRISE AND SUNSET

At sunrise and sunset, the light from the Sun has to pass through quite a large part of the atmosphere and consequently, the sky itself goes through an often beautiful sequence of colours. When the Sun is low on the horizon, its light is passing through the densest part of the atmosphere. The air scatters all the shorter wavelengths of light aside, so that the Sun appears a deep golden yellow or red. Just above the position where the Sun has set, or is just about to rise, is known as the twilight arch.

At sunset, as the Sun sets over the horizon, above the point where the Sun was last visible the sky at the horizon is a pale yellow, with a bluish-white coloured segment above it. On either side, the sky is orange at the horizon, shading upwards to a pale yellow. Keep watching and you'll see the twilight arch become orange at the bottom, graduating into yellow and higher in the sky, a gorgeous pink. Sometimes there is also a slight green tinge on either side where the yellow area graduates into the blue of the sky. In time, the pink area will 'sink' towards the horizon, flattening out as it 'falls' while the sky above changes from a blue-grey to purple-blue and, eventually, into the deep blue of the night sky. Along the horizon line, the last 'specks' of light are often a yellowish green. At sunrise, the sequence of colours is reversed and because there is often less haze (which builds up through the day), the colours are frequently stronger – so it's worth getting out of bed extra early to see them!

Purple light and green flashes

On very rare occasions, the top of the twilight arch may become the most incredible vibrant purple – much more intense than the 'purple tint' that is more usually seen at twilight. The intense purple light is very short-lived – only on very few occasions has it been reported as lasting more than a few minutes – and, because of the properties of colour films, it is one of the optical phenomena that is very hard to .

RIGHT: *Beautiful cloud colours at sunset are among the most common optical phenomena.*

photograph, although advances in digital technology may overcome this. The purple light is rare because it is only normally seen after there has been a substantial volcanic eruption which has sent debris such as fine dust or sulphur dioxide particles high into the stratosphere. The particles are large enough to scatter significant amounts of long-wavelength red light from the Sun: the purple light is produced by mixing this red light with the ever-present blue light scattered by oxygen and nitrogen molecules in the air.

The famous Green Flash sometimes occurs when the air is very clear. When the Sun sets below the distant horizon, the last little bit of Sun that is visible very occasionally appears as a flash of brilliant green light. On even more rare occasions it appears as a flash of blue-violet light. Sometimes the effect can last a little longer than a 'flash': the last segment of the setting Sun can appear green, which is called the Green Segment. There is some debate about the nature of the Green Flash and Green Segment. Some people believe that it is an optical illusion: staring at the Sun (which is dangerous and should be avoided!) means that our eyes become insensitive to red light and we experience a shift in apparent colours, so what we perceive as green is in fact yellow. However, the Green

Flash is sometimes visible at sunrise, when there has been no previous staring at the Sun!

As the Sun sinks slowly in the west...
At sunset and at sunrise, high cumulonimbus clouds and the tops of mountains are sometimes magnificently lit by glorious sequences of colours. This is known as Alpine Glow because it was first noticed on the snow-capped Alps of western Europe. Alpine Glow is generally strongest when the weather is fine and the air is clear. As the shorter wavelengths of light (the blue and green wavelengths) are scattered by the atmosphere, only the longer wavelengths (red, orange and yellow) remain to travel on their path through the atmosphere. The mountain peaks become tinted with a range of colours in sequence: yellow, pink, red, and finally purple – because some scattered blue light does nevertheless reach the Earth! At sunrise, the sequence of colours is reversed.

Aurorae
The Aurora Borealis, or Northern Lights and its cousin in the southern hemisphere, the Aurora Australis, don't have a direct relationship with the weather but are included here because the gorgeous displays of light can be seen at twilight and at night. The greatest frequency of aurorae occurs in the region known as the auroral zone which lies around 15°-30° from each magnetic pole. This means that the Aurora Borealis can frequently be seen in Northern Europe, Scandinavia, Iceland, Siberia, Canada and Alaska in wintertime, but are invisible during the long summer twilight. Very infrequently – and the cause of much excitement in lower latitudes – major aurorae displays are seen as far down as the geomagnetic Equator. On 13 March 1989 a magnificent display was visible across Great Britain and northern France – and observers were equally amazed to see the display to the south of them, rather than low on the northern horizon!

ABOVE: *The Aurora Borealis are best seen in northern latitudes.*

The luminous phenomena of aurora (which is the Latin for 'dawn') occur in the upper atmosphere at altitudes of between 100 and 1000 km (60 and 620 miles) in the mesosphere and

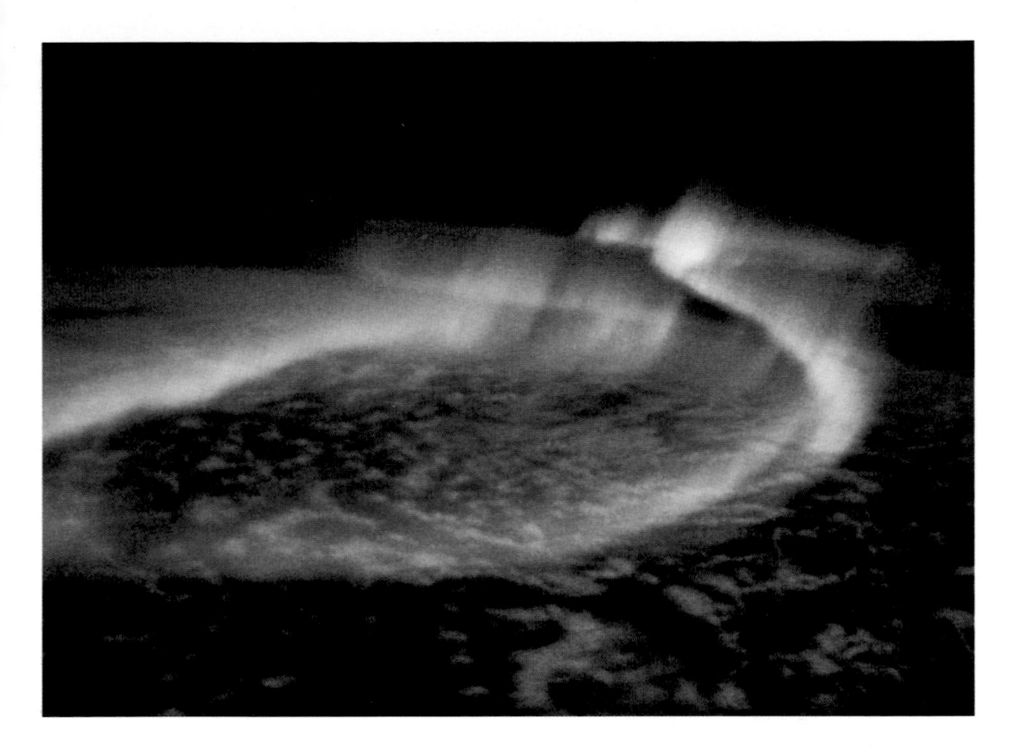

ABOVE: *Southern lights.*

thermosphere, when energetic particles – protons and electrons – from the Sun enter the Earth's magnetosphere and 'excite' the atoms in it. It's this 'magnetic activity' that, as well as producing aurorae, makes mariners' compasses useless! Collisions between the electrons and atmospheric oxygen and nitrogen cause the 'excited' molecules to emit energy in the form of different coloured light: what is called 'auroral green' is emitted from the excited oxygen, while red light is emitted from the 'excited' nitrogen. Auroral rays at the top of a display may in fact be at about 1000 km (620 miles) in altitude, high up in the thermosphere at the edge of space and are therefore in sunlight where ionised nitrogen molecules emit a bluish-violet or purple light. However, the colours that are actually seen really depend on the strength of the aurora, and on each person's eyesight. Weak aurora can appear pale green or even colourless, and people who have poor low-light sensitivity to red may be unable to see the red or purple tints, even in very strong displays!

An auroral display may begin with a faint 'glow ' on the horizon towards the pole: a 'veil' – a weak but even glow of light covering a large area of sky, or a 'patch' – a weak but luminous area, often looking a little bit like glowing cumulus cloud, but which fluctuates in intensity. The display then most commonly assumes the form of an 'arc', an arch-like structure with a sharply defined lower edge but a diffused upper border. From this, the aurora may develop 'rays' – almost vertical streaks of light. Strong aurora displays may also develop multiple 'bands' which are deep, ribbon-like structures often described as being like 'folded curtains'. If the display moves quite rapidly overhead, the rayed structure may appear as a corona – where the rays seem to radiate from a point high overhead. No two aurorae displays are the same, and their intensity may vary slowly over a period of a few minutes, a variation known as 'pulsations'; there may be more rapid 'flickering' affecting all or parts of the display, or there may be 'flaming' – a great surge of brightness that sweeps up from the horizon. Towards dawn, the brilliant display breaks up into pulsating patches of light before the Sun once again takes command of the sky.

RIGHT: *In auroral displays, collisions between electrons and atmospheric oxygen and nitrogen, cause the molecules to become 'excited' and emit energy in the form of different coloured light.*

LIGHTNING

WHILE THUNDER IS THE weather's most audible phenomena, lightning must be its most dazzling! Lightning is an enormous electric spark between regions of oppositely charged particles. Lightning develops inside growing cumulonimbus clouds as negatively charged electrons migrate from freezing ice crystals to liquid water droplets, resulting in an overwhelmingly negative charge in lower portions of the clouds with positively charged areas in the still icy cloud areas above. Since like charges repel each other, the negative charges in the lower part of the cloud push the electrons on the ground outwards to the edge of the storm, leaving a residual positively charged area directly beneath the cloud. The voltages build up in all the regions of opposite charge, and when they are large enough, electric discharges – lightning – occur between the opposite-charged areas. In fact, most lightning activity takes place inside the cloud itself: this is called 'in-cloud lightning', and only about 20% of discharges are of the 'cloud-to-ground' variety.

We often speak of 'forked lightning' and 'sheet lightning', but the distinction is somewhat artificial: forked lightning is where the vertical channel of electric discharge can be seen, whereas with sheet lightning, the channel is often invisible. This is because the discharge channel is often obscured by clouds – especially when flashes occur in different parts of a single cloud (in-cloud lightning) or between clouds, known as 'cloud-to-cloud lightning'.

The familiar 'cloud-to-ground lightning' begins with a negative discharge from the cloud, known as a 'leader'. When the

LEFT: *A spectacular storm over Almeria, Spain.*

RIGHT: *Cloud-to-ground lightning.*

leader approaches the ground, an upward flow of positive electrical energy, known as the 'return stroke', rushes upwards to meet it. There may be several return strokes, but to our eyes they appear as one, what is called 'streak lightning'. (If the light channel is branched, it is then called forked lightning.) When the lightning flash is so distant that we see only the sky lit up above the storm – but don't hear the thunder – then it is known as 'heat lightning'.

Light travels much faster than sound: 300,000 km (186,000 miles) per second compared to a rather sluggish 340 m (1,110 feet) per second for sound waves! So, it takes about 3 seconds for thunder to travel 1 km (or 5 seconds per mile), but when lightning flashes, you see it straight away! To work out the number of kilometres between you and a thunderstorm, count the number of seconds between the lightning flash and the arrival of the sound of thunder, and then divide by three (or five, for miles).

GOODNESS, GRACIOUS, GREAT BALLS OF FIRE!

There are some other unusual, and rare, forms of atmospheric electricity which, when seen, should be reported as they are of great scientific interest! Ball lightning is a rare phenomenon which has been observed and described as a luminous sphere about 15 cm (6 in) in diameter that appears to persist for a few seconds after a close cloud-to-ground discharge. Sometimes the ball 'rolls' along and 'jumps' or 'bounces' about before it 'fizzles' out – rarely does ball lightning explode! It's not certain how ball lightning is formed but scientists would really like to know as it may help them to develop new means of energy production and transmission! Bead lightning is so-called because it appears rather like beads or pearls on a string – it is sometimes called 'pearl-necklace'

LEFT: *Light – and the visible lightning we see – travels at 300,000 km (186,000 miles per second.*

ABOVE: *Forked lightning is so-called because the channels of electric discharge are visible.*

lightning. This is a very rare form in which variations of brightness occur along the discharge channel that produces a transient appearance.

Rocket lightning

For many years rocket lightning was thought to be an optical illusion. It wasn't until the advent of supersonic jet flights and orbiting Space Shuttles that this phenomenon was verified because the tops of storm-making cumulonimbus clouds reach to the edge of the troposphere at over 10 km (6 miles) in altitude. Rocket lightning is very slow, so slow it can be followed with the eye! The only problem is, it projects upwards from the top of thunderstorms into clear air above, which is why it has most often been observed by the 'jet set'!

St Elmo's Fire

St Elmo's Fire is a luminous glow which can be seen on objects 'on the ground' such as weather vanes, chimney pots, the tips of trees and shrubs – even on the tips of someone's hair! It is also seen at sea on the tips of spars and masts, and in the air on the tips of airplane wings. Technically, St Elmo's Fire is a corona discharge (or a brush discharge) from the tips of sharp objects and is often accompanied by a 'fizzing' or crackling sound. The name is derived from the patron saint of (Italian) sailors, St Ermo (died AD303, also known as Saint

BACKGROUND IMAGE: *Thunderstorm.*

NEXT PAGE: *A spectacular display of lightning bolts over a city.*

Ulmo and Saint Erasmus in various other parts of Europe), whose 'appearance' on board sailing ships was believed to bring protection in storms! St Elmo's Fire is also known as 'corposant' from the Portuguese *corpo-santo* (holy body).

LIGHTNING SAFETY

Although only about one in five lightning strokes are from the clouds to the ground, each one instantly heats the channel of air through which it flows by about 30,000°C (54,000°F)! The violence, speed and danger of lightning have captured human imaginations for centuries. While the ancient Greeks believed that Zeus was giving vent to his anger by hurling thunderbolts from Mount Olympus, the Romans associated lightning with the god Jupiter. 'Pagan' or pre-Christian Britons believed that acorns offered magic protection: they can still be seen as decorative finials on gate posts or as door knockers on many British houses! Early Christians put their faith in St Donatus, a miraculous survivor of a lightning strike. In many representations of the saint, he appears dressed in Roman armour, carrying a sword – not a good idea in a lightning storm – and carrying a 'protective' palm leaf. In fact, the only device that has shown itself to be completely and consistently effective in controlling the weather is Benjamin Franklin's lightning rod which he invented in 1752, following his experiment with a kite and key during which he proved the electrical nature of lightning. (As additional insurance Franklin was also responsible for establishing the United States' public Fire Service.)

If you are outdoors when lightning occurs, avoid open spaces where you are the highest object, and take cover in a building or car. Don't stand

RIGHT: *The lightning rod has proved effective in controlling the power of lightning strikes.*

ABOVE: *Each strike heats the channel of air it passes through to about 30,000°C (54,000°F)*

underneath an isolated tree – if this is the tallest object around, it may be struck and subsequently injure you. If there is no cover available, find the lowest spot you can – go into a valley rather than up a hill, or even a dry ditch. Don't lie flat on the ground, but crouch with your arms around your legs and your head on your knees. If possible, take off any metal jewellery, watches and metal-framed spectacles: if you do get struck by lightning, you will probably survive the strike, but can be seriously burned by molten metal! St Donatus was blessed – or just very lucky!

LEFT: *Lightning will often strike isolated trees as it tries to find the quickest way down to the ground.*

If you are indoors when lightning occurs, stay well away from windows, water pipes, telephones and electrical wiring. Switch off and, if possible, disconnect electrical appliances including phones and computers.

MIRAGES

MOST PEOPLE ASSOCIATE mirages with hot, dry deserts – and thirsty, wandering travellers lured by illusions of an oasis! Mirages, however, are not confined to hot deserts, but can be seen at the poles, and even along stretches of roads on hot days. Mirages are caused by refraction – light deviated from its original path – in the lowest layers of the Earth's atmosphere, which affects the image of distant objects. When there are very steep temperature gradients in the atmosphere the air at different temperatures has different densities, and it is the variations in the densities which determine the degree to which the light is refracted. Depending on the density profile of the air, images may either be compressed or elongated vertically, appear to be elevated above or 'depressed' below the horizon. The image can also appear upside-down, or even split into multiple layers.

INFERIOR AND SUPERIOR MIRAGES

The two main types of mirage are inferior – where objects appear lower than normally expected, and superior – when they appear higher than expected. The pools of water that appear to be lying on the road on hot

RIGHT: *A mirage on a desert highway: this is a common example of an inferior mirage.*

ABOVE: *Fata Morgana is responsible for almost mystical images to appear in the sky – in reality distorted reflections of objects on the surface.*

days are a form of inferior mirage: the 'water' is really an image of the sky appearing lower than expected. The mirage occurs because the lowermost layer of air nearest the ground has become very strongly heated and therefore, less dense than the air lying over it. The rays of light from the Sun are refracted or bent upwards, giving the appearance of being reflected from the hot road surface.

Superior mirages are less frequent than inferior ones, but they can be seen over the sea and where there are large expanses of snow and ice. A superior mirage occurs when the lowest layer of air is colder than the layer lying over it, such as when warm air crosses a cold area of sea, or an ice- or snow-covered surface. Rays of light from distant objects are refracted downwards and reach the eye along paths which suggest to our brains that the objects come from the sky – they seem to float in the air above the horizon.

Fata Morgana

Superior mirages may produce multiple images that can be both upright and upside-down. A very striking effect of a superior mirage is a Fata Morgana. The name comes from the Italian for 'Morgan the Fairy', the half-sister of the legendary King Arthur, who was reputedly able to create such illusions. In a Fata Morgana, small surface details are distorted, inverted, and elongated vertically – or can be stretched horizontally. One or more temperature inversions are at work here to create the appearance of 'castles in the air': distant objects appear elongated giving the impression of castle turrets, spires, tall buildings and towns in the distance.

Somewhere Over the Rain-moon-fog-dew-bow

When the weather forecaster says that the day will be one of 'sunshine and showers', then chances are there will be rainbows. Although they are common – most people will have seen one – they are always a welcome sight! After 40 days and 40 nights of non-stop rain, Noah in his Ark knew that God had stopped the flood because he sent a rainbow. To the Romans, the rainbow was personified as the goddess Iris, who brought messages from the gods to mortals on Earth. In the 13th century, many believed that walking under a rainbow could cause you to change sex! Meanwhile, at the end of every rainbow in Ireland is a pot of gold guarded by a leprechaun.

We all know the colours of the spectrum or rainbow: red, orange, yellow, green, blue, indigo and violet. It was Isaac Newton who discovered that sunlight, or white light, was composed of all the colours of the spectrum. Using a ray of sunlight directed through a glass prism, Newton noticed how the ray of light was bent or refracted, splitting the white light into the different colours. If all the colours

ABOVE: *A common, but nevertheless pleasing, sight, rainbows accompany days of sunshine and showers.*

ABOVE: *Rainbows appear in the anti-solar part of the sky – opposite the Sun.*

are projected together, for example by a lighting engineer in a theatre merging spotlights of these colours together, they create white light.

Newton also discovered that each colour has a different wavelength. Red has the longest wavelength, while violet has the shortest. Because each colour has its own wavelength, we see different colours when light is reflected off the surface of an object: a red table looks red because it absorbs all of the wavelengths of light except red, which is reflected or 'bounced back' to our eyes. Light rays can be refracted as they enter a droplet of water: inside the water droplet they reflect off the far side and are further refracted as they leave. The rays of light exiting from a raindrop are returned at an angle of about 42° from the incoming rays. The result is a rainbow.

Because the light is reflected inside the raindrop, rainbows appear in the side of the sky which is opposite the Sun –the anti-solar point. That point is determined by

BELOW: *White light refracted through a prism splits it into its component spectral colours.*

the position of the viewer, so each person sees their own individual rainbow! As you move, so does the rainbow – sadly there's never any chance of getting to the pot of gold!

The amount of the circular arc that is visible depends on a number of factors. The height of the centre of the arc depends on the altitude of the Sun: the higher the Sun in the sky, the lower the rainbow; the lower the Sun, the higher the rainbow. A 'perfect' rainbow appears to stretch right across the sky and 'touch' the ground at both ends: this perfect semicircle occurs when the Sun is on the horizon. Sometimes, from a very high vantage point, like the top of a mountain or from an airplane, a complete circular rainbow is visible. Most of the time only one side or the other of the rainbow can be seen: this happens because only some of the raindrops are illuminated by sunlight and the rest are in shadow, or because it is raining in only part of the sky and there are no raindrops to form the rest of the bow. The strength of colours in a rainbow depends on the size of the raindrops: the bigger the drops the stronger the colour.

ABOVE: *The strength of the colours in a rainbow depend on the size of the raindrops: the bigger the drops, the stronger the colours.*

ABOVE: *In a full arc rainbow, both ends appear to touch the ground*

PRIMARY AND SECONDARY RAINBOWS

The most commonly seen rainbows are known as primary bows and have a radius of around 42°. The spectral colours always appear with red on the outside edge and violet

ABOVE: *In a secondary rainbow, the sequence of colours is reversed.*

on the inside edge. Sometimes, though, light is reflected twice by the raindrops before being sent on its way back towards our eyes: the first set of raindrops produces a primary rainbow, while the second produces a secondary, and much larger rainbow (with a radius of about 51°) which appears on the outside of the primary bow and in which the sequence of colours, although fainter, is reversed. The area of sky in between the two bows is much darker than the surrounding sky: this area is known as Alexander's Dark Band and is darker because the raindrops in this area are sending the light in a different direction, not towards our eyes. Under certain conditions, it's possible to see one or more bows lying 'underneath' –

inside – the primary bow. These rather pale-coloured bows are called superanuary bows, or sometimes, interference bows, and they occur when light takes paths of slightly different lengths through raindrops.

MOONBOWS, DEWBOWS AND FOGBOWS

The Moon also produces 'rainbows', but because its light is not as strong as that of the Sun, the colours in a moonbow are generally very pale and most of the time appear as white. Moonlight is reflected sunlight, so all the colours of the spectrum are actually present. Moonbows are most frequently sighted around tropical islands where night-time rain showers and partly cloudy skies are more common than at higher latitudes. Bows identical to rainbows also occur with water droplets from fountains, hose pipes and of course, waterfalls: the spectacular Niagara

Falls on the Canada/USA border are frequently adorned by 'nature's paint box'.

Droplets of dew act in the same way as raindrops and produce a coloured bow, a dewbow, which can often be seen on grass-covered areas in autumn. However, dewbows don't appear as circular arches because only one or two portions are visible on a horizontal surface such as a lawn, lying to the left and right of the anti-solar point. Thanks to spiders who overnight have been busy spinning hundreds of tiny webs between the blades of grass, the dewdrops have been suspended above the ground and act just like raindrops in the sky.

With very fine water droplets, such as in mist and fog, the light is not reflected and refracted inside the droplets, but is diffracted by them instead. Consequently nearly all the colour disappears and a white fogbow can be seen.

Glory

A glory consists of a series of coloured rings which appear around the anti-solar point when sunlight is falling on a cloud or a bank of mist or fog. Blue-violet appears on the inside while red appears on the outer edge. Sometimes, more than

LEFT: *Water droplets from waterfalls also create spectacular rainbows.*

ABOVE: *Fog bows are created by very fine water droplets such as mist and fog. The light is not refracted, but defracted, so the rainbow colours disappear, leaving a white fog bow instead.*

one set of rings is visible. The most common 'sighting' of glories is from airplane windows where they appear surrounding the shadow cast by the aircraft on a cloud bank.

Heiligenschein

Meaning 'holy light' in German, a heiligenschein is a bright colourless 'halo' of light that appears around the shadow of the observer's head – that is, around the anti-solar point. Strangely, you can see only the light surrounding the shadow of your own head – not that surrounding any companions! There are two factors at work here: sunlight falling on dew-covered grass (the dewdrops and grass act as 'retro-reflectors', returning the light preferentially towards the source in the same way that a 'cat's eye' works on roads), and a rough surface, such as dry grass. When you look, you don't see the shadows of the blades of grass, but instead every visible surface is illuminated, so the immediate area around the 'shadow' of your head appears brighter – like a halo. The bright spot of the heiligenschein is often visible from aircraft, aerial photographers call it the 'hot spot', and it can be seen gliding across fields below. Next time you watch TV with footage taken from aircraft – look out for glories and heiligenschein!

Optical Phenomena Near the Sun or Moon

RAINBOWS, AND OTHER BOWS, as well as glories and heiligenschein, all occur in the sky opposite the Sun (or Moon), in the position known as the anti-solar point. Other optical phenomena occur near or surrounding the Sun (or Moon). It is vital that care is taken when viewing these phenomena as looking directly at the Sun can cause irreparable eye damage.

CORONAE

A corona (the Latin for 'crown') is an optical phenomenon that consists of one or more sets of coloured rings surrounding the Sun or Moon. (It should not be confused with the corona which is the extremely hot outer surface of the Sun!) The rings are produced by diffraction, deviation of light from its original path, by water droplets in thin stratiform cloud. The colour and strength of the colours depend on the size and uniformity of the water droplets in the clouds: the more uniform the particle sizes, the purer the colours.

Coronae consist of an inner aureole which appears as a bluish-white disk around the Sun or Moon, with a brownish-red outer edge. If the cloud particles are of different

KRAKATAU AND THE REVEREND BISHOP

In 1883 Krakatau, a small volcanic Indonesian island lying between Java and Sumatra, famously erupted throwing thousands of tonnes of volcanic ash, dust and sulphur dioxide into the upper atmosphere. The Reverend Bishop reported from Hawaii that the Sun was later surrounded by a pale disk of light with a faint reddish-brown outer zone which had an inner radius of around 10° and an outer radius of around 20° when the Sun was at its highest point. The diffraction in this case was caused not by cloud droplets but by the tiny particles of volcanic debris lifted into the atmosphere by the volcanic eruption. Since 1883, the Bishop's Ring has been reported following other volcanic eruptions.

sizes rather than being uniform, it is often the case that the aureole is the only part of the corona that can be seen. A full corona will show an aureole and an outer set – or sets – of rings, ranging in colour from violet on the inside to red on the outside. The radius of the rings is also dependent on the size of particles in the clouds: the smaller the particles, the larger the radii of the rings! Because the particles in the clouds may be unevenly distributed, a corona may be irregular in its outline: indeed coronae tend to change shape and colour as clouds pass across the Sun or Moon.

ABOVE: *Halo around Moon*

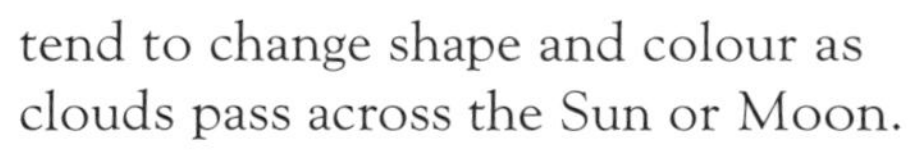

HALOES

'Ring around the Moon, Rain is coming soon' is an old English saying – and a fairly accurate weather observation too! The 'rings' are haloes, circular arcs around the Sun (or Moon), and they are extremely common in Britain and western Europe. They occur whenever there is a thin veil of cirrostratus cloud which often precedes a warm front (and rain). The light is bent and scattered as it passes through two non-adjacent faces of a hexagonal ice crystal (with an included angle of 60°) in the clouds to form a halo, most frequently with a radius of 22°. The ice crystals are usually random in their orientation, so the light is spread in all directions, and the halo may be incomplete and vary in intensity. When faint the halo appears white, but when strong the inner edge can appear red, surrounded by yellow.

There are many halo phenomena: parhelic circles ('mock' Sun rings _ a white circle that passes through the Sun and lies parallel to the horizon); circumzenithal arcs (bands of brilliant spectral colours centred on the zenith and convex towards the Sun); circumhorizontal arcs (bands of brilliant spectral colour which run parallel to the horizon below the Sun in the sky). The appearance of all of these phenomena is governed by the shapes of the ice crystals in the clouds, the elevation of the Sun, and the degree of latitude of the observer. Other, more rare halo phenomena are still little understood, but of great interest to the meteorological community and sky watchers worldwide!

NEXT PAGE: *A corona around the Sun is produced by light deviated from its original path by water droplets in thin stratiform cloud.*

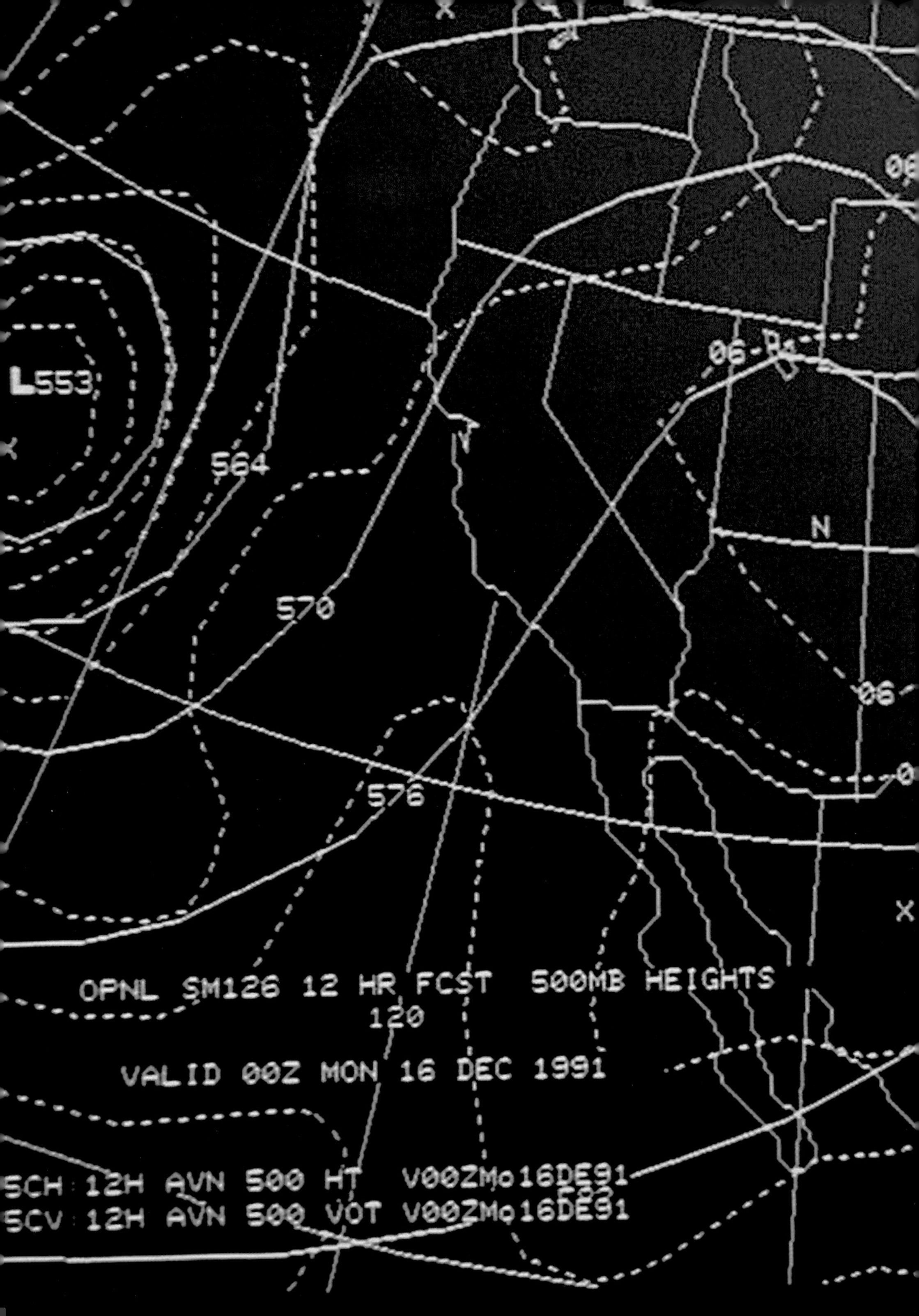

L553
564
570
576
06
OPNL SM126 12 HR FCST 500MB HEIGHTS
120
VALID 00Z MON 16 DEC 1991
SCH 12H AVN 500 HT V00ZMo16DE91
SCV 12H AVN 500 VOT V00ZMo16DE91

MAPPING AND FORECASTING

FOR CENTURIES, HUMANS HAVE TRIED to predict the weather by observing the current weather and the behaviour of animals and plants. Often expressed as rhymes, these observations became a sort of folk-science and, in some instances, could be a very accurate prediction of immediately impending weather: 'When goose and gander begin to meander; The matter is plain: they are dancing for rain!' , 'Swallows fly high: clear blue sky; Swallows fly low: rain we shall know' and 'When bees stay at home, rain will soon come; If they fly away, fine will be the day'. Such predictions can work – but only over short time-scales. Less accurate are Groundhog Day and St Swithin's Day.

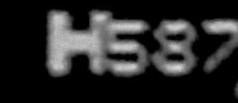

GROUNDHOG DAY (2 February) takes place at Punxatawney, Pennsylvania, and involves a captive groundhog known as Punxatawney Phil. If 'Phil' sees his shadow when he pokes his head out of his burrow, then winter will (supposedly) last another six weeks; if he doesn't see his shadow, the spring is just around the corner! That corner however, could take six weeks to turn! St Swithin's Day (15 July) is named after Swithin, Bishop of Winchester who died in AD862 and was buried at his own request in the grounds of the cathedral. In the next century, Swithin was formally canonised as a saint and the decision was made to move his remains to the choir inside the cathedral on 15 July. The plan was abandoned, however, after 40 days of non-stop rain which began on that day. In due course, the event passed into folklore. In neither case, though, can the weather on one particular day be an accurate prediction of the weather to come. Even with the advent of supercomputers and orbiting satellite technologies, meteorologists still find it difficult to forecast the weather 40 days in advance.

The ingredients of a weather forecast are the observations of the weather which are made simultaneously around the world. In order to predict what will happen to the weather over the next few hours, days, weeks or even a month ahead, it is vital that meteorologists can measure what is happening now.

ABOVE: *Despite folklore, one days' weather cannot accurately predict the weather for the coming weeks.*

ABOVE: *Groundhog Day: if Punxatawney Phil, the groundhog, sees his shadow on 2 February, winter will last another six weeks in Pennsylvania.*

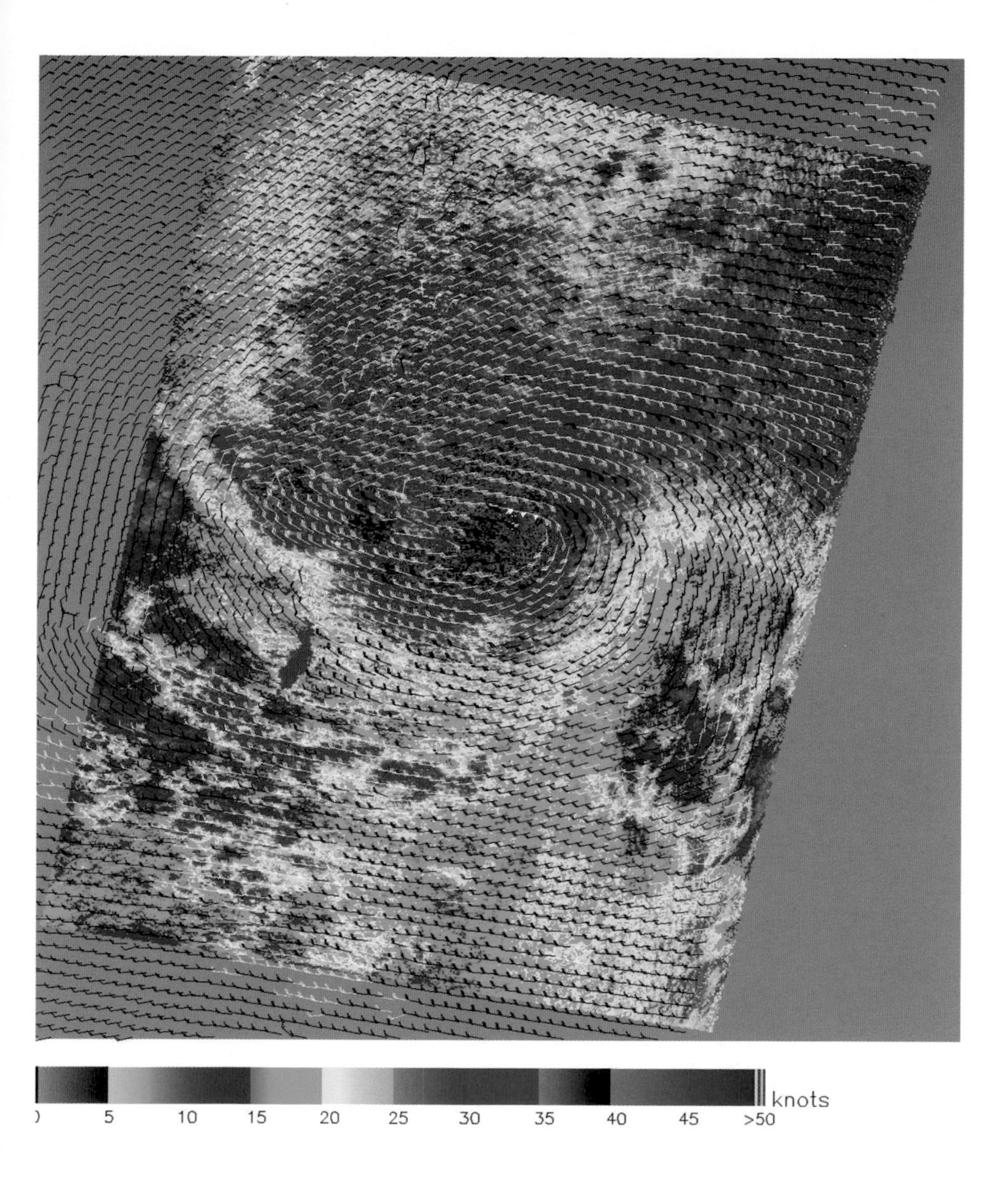

ABOVE: *Satellite technology: produced from data collected by the SeaWinds scatterometer instrument onboard NASA's QuikSCAT mission reveals the details of the surface winds and rain in Typhoon Nanmadol as it moves westward. The data was collected on 1 Dec. 2004 at approximately 8 in the morning.*

Taking Measures

MEASURING THE WEATHER is done in a number of ways. Surface observations are made at weather stations across the globe, some of which are manned, while others are fully automated.

TEMPERATURES: DRY AND WET BULB TEMPERATURES

Air temperature is measured with a mercury-in-glass thermometer which is read to the nearest 0.1°C. This is known as a dry bulb thermometer and at weather stations it is housed (along with other instruments) inside a weather screen. The screen ensures that the thermometer measures air temperature – that is, the temperature of the air flowing through the screen via the gaps between the downward-angled slats on the side of the box. The outer box housing the equipment is painted a reflective white and has an insulated roof and floor to ensure that the air temperature being read has not been 'interfered' with or warmed up or cooled down by sunshine, or the ground temperature below the screen.

A second thermometer is also housed in the screen, but its bulb is

SURFACE OBSERVATIONS

The usual surface observations, which are reported every hour, are:

dry bulb temperature
dew point temperature
mean-sea-level barometric pressure
pressure tendency
total cloud amount
cloud type and base height
horizontal visibility
wind direction and speed
present and past weather
precipitation total (usually a 12- or 24-hour period total)

covered by a muslin bag which is kept permanently wet with distilled water supplied by a wick. This is the wet bulb thermometer, and while it measures the wet bulb temperature in degrees C or F, it is in fact a measure of humidity.

The wet bulb thermometer is the instrument by which meteorologists calculate absolute and relative humidity. Absolute humidity means the maximum amount of water vapour in grams that can be contained in a cubic metre of the air and water vapour mix. Relative humidity refers to the ratio (expressed as a percentage) of the actual amount of water vapour contained in the sample of air to the amount it could contain if saturated at the observed dry bulb temperature. The specific humidity (the mass of water vapour in grams present in a kilogram of dry air) of air that is saturated with water vapour increases with temperature: saturated air with a temperature of 0°C contains 3.0g/kg; at 10°C this rises to about 7g/kg. At 20°C it's about 14.0g/kg, and at 30°C it's 26.0 g/kg.

As well as 'spot values' of humidity, a time trace of fluctuations in relative humidity is often given by a recording hygrometer. This makes use of the fact that a horse, or human, hair lengthens or shortens as relative humidity varies. Human hair shrinks by about 2.5% when relative humidity reduces from 100% to 0%.

In the recording hygrometer, a small sheaf of hair is stretched across a thin metal bar which is connected mechanically to a pen that records the fluctuations in humidity on a strip chart wrapped around a rotating drum. The strip chart is usually changed once a week. Other hygrometers use the moisture absorbing properties of various chemicals, which become drier or moister as the humidity increases or decreases. These machines are the scientific equivalent of seaweed, a piece of which hung outdoors has for a long time been used as an indicator of relative humidity!

ABOVE: *Typical outdoor thermometer.*

Maximum and minimum temperatures

Inside the screen are also horizontally mounted maximum and minimum thermometers which are designed to record the highest and lowest temperatures that occur during a specified period – often 24 hours from 0900 local time. Usually, the maximum temperature occurs mid-afternoon, and the minimum temperature during the hours of

LEFT: *Windsocks are used at aerodromes so pilots can gauge wind speed and direction quickly.*

darkness. The minimum thermometer contains alcohol instead of mercury because it has a much lower freezing point (-114.4°C compared to mercury's -38.9°C), making it far more useful in extremely cold regions of the world. When a weather report forecasts 'tonight's low' or 'today's high', it is the minimum thermometer reading to which it is referring.

THERMOGRAPH

The thermograph, though a little old-fashioned, is still widely used at main weather stations. Unlike thermometers which give a 'point value', thermographs provide a continuous pen trace of the temperature, usually over a week. A thermograph is basically a pen on the end of a long 'arm' which is attached to a coil that distorts as the air temperature rises and falls. The pen then traces the fluctuations on a thermogram – a paper chart wrapped around a clockwork-driven drum. In automated weather stations, more modern electrically operated machines are used.

PRESSURE AND BAROMETERS

Measuring atmospheric pressure is to weigh the air that presses down on the Earth. Pressure decreases with altitude throughout the atmosphere because there is progressively less air above a given level.

Air pressure can support a column of water – or other fluid – in a glass tube immersed in a reservoir at its lower, open end and topped by a vacuum at its upper, sealed end. At sea level, atmospheric pressure is such that the column of water would be about 10 m (33 ft) high, which is a bit big for a weather station. Because of the density of mercury however, the height of the column is a much more manageable 75 cm (30 in).

The mercury barometer was developed in the 1640s by Evangelista Torricelli (see page 20), a student of Galileo, and is still widely used today. However, the readings must be adjusted to take account of the surrounding air pressure and gravity. Station level pressure is read to the nearest 0.1 mbar (0.1 hectopascal or hPa), but this reading too needs adjusting to a common datum, mean-sea-level. This means that a certain number of millibars is added to represent the

LEFT: *Low pressure on a storm glass barometer indicating bad weather to arrive shortly.*

NEXT PAGE: *A meteorological station.*

pressure of an imaginary column of air between the barometer and the mean-sea-level. If the weather station is below sea level, then the adjustment is made by subtracting a number of millibars from the reading.

An aneroid, or 'without air', barometer is a familiar device that can be seen in many homes. This type of barometer senses pressure through small distortions of a partially evacuated metal capsule. Higher atmospheric pressure will 'squash' the capsule more than lower atmospheric pressure. The capsule is linked to an arrow that moves around the scale of millimetres and/or inches of mercury, and of millibars. These readings are also often accompanied by decorative, though not particularly reliable 'text' forecasts of 'Dry', 'Wet', and 'Change'.

Barograph

Like the spot temperatures in thermometers, the problem with barometers is that they provide only an indication of the air pressure at the time at which they are read. This makes the barometer somewhat limited in its application. More significant to meteorologists is the rate of change of pressure over a particular time at each station, and any patterns of pressure across the surface at mean-sea-level. Change of pressure with time is recorded with a barograph, an aneroid instrument with a mechanical arm that traces a continuous line of pressure on a barogram – a thin paper strip that is changed weekly. The barograph is more useful because it allows meteorologists to see the size and direction, up or down, of a station's pressure change and, typically, this covers a three-hour period leading up to the actual observation time.

PRECIPITATION

The most common gauge used to measure precipitation is a rain gauge that is emptied once a day to give a simple record of fall in millimetres or inches. The gauge is often a 12.7-cm (5-in) diameter copper cylinder with its top 30.5 cm (12 in) above the surrounding surface in order to minimise the risk of water splashing off the ground. Once a day, the water is decanted into a tapered measuring glass to determine the level of precipitation to the nearest 0.1 mm or 0.01 inch.

Once again, daily tallies are useful, but they can't tell us about intensity or duration. 'Auto graphic' recording devices have been designed to record such details as a paper 'hard copy' or as a tele-metered radio message from the gauge to the central weather service.

Additionally, precipitation radars are used to produce maps: the radar makes a circular scan every 15 minutes and emits pulses of radiation. Small fractions of the radiation will be reflected back to antenna by precipitation size particles – not cloud droplets which are much smaller. The radar then converts the reflected radiation into an image showing the extent and intensity of precipitation as it falls to Earth.

ABOVE: *An anemometer measures wind speed.*

CLOUD TYPE AND AMOUNT

Cloud cover is reported as oktas or eighths of the sky covered, as both individual layers of cloud and total cloud amount (see page 133). This summarises one or more cloud types that may be present. Clouds are allocated to one of three layers: low, middle and high. The layer in which they occur depends on the height of their base above the surface of the Earth.

WIND SPEED AND DIRECTION

Wind speed is measured by an anemometer. This usually has three cups in the form of hemispheres which are mounted on to a vertical shaft. The pressure of the wind on the concave inner faces of the cups is greater than that on their convex outer faces, and this causes the vertical shaft to rotate. The rate of rotation varies with wind speed which is displayed on a calibrated dial marked with knots (nautical miles per hour) and metres per second. Combined with the anemometer is a wind vane that points into the wind, indicating the direction from which it is blowing. A wind direction report is usually given as an average taken over a few minutes and is expressed in degrees read clockwise from true north to the nearest ten degrees. 000° is reserved for calm conditions when there is no wind; an easterly wind (that is, a wind blowing from the east) has a direction of 090°; southerlies blow from 180°; westerlies from 270° and northerlies from 360°. In-between are finer gradations: a southwesterly would be 225° on the scale.

RIGHT: *A weather station with precipitation radar, anemometer and wind vane.*

The duration of 'bright sunshine' is measured on a piece of sensitised card held in a frame and wrapped around one half of a glass sphere that focuses the Sun's rays on to it. The term 'bright' is a clue to the fact that the recorder isn't sensitive enough to record sunshine at dawn and dusk, it only provides a total duration of bright sunlight to the nearest tenth of an hour every day. Some observation sites are equipped with solarimeters which measure the intensity of solar radiation (expressed in watts per square metre) received on the surface of the Earth. Solar radiation takes two forms: direct radiation which reaches the instrument directly from the Sun, and diffuse radiation (or sky radiation) which arrives after it has been scattered by gas molecules, dust and other particles in the atmosphere.

Knowing about bad visibility is more important than knowing about

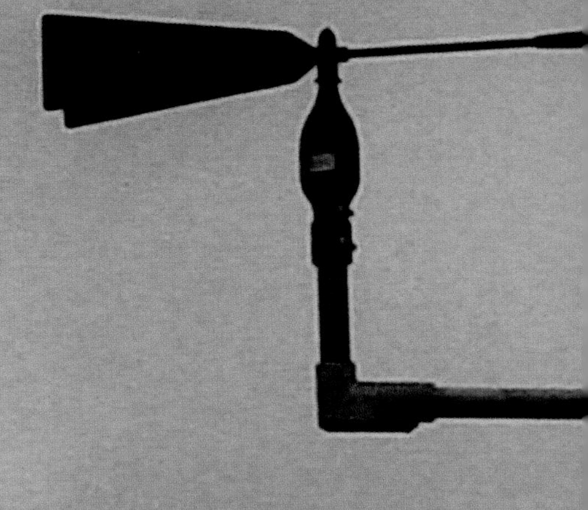

LEFT: *Wind vane and anemometer.*

good visibility, as poor visibility can be a dangerous hazard to traffic, both on the roads and in the air, and to shipping. On land, visibility is quoted to the nearest 100 m up to 5 km, then to the nearest kilometre from 5 km to 30 km, and then every 5 km up to a maximum of 75 km. When visibility is extremely poor, less than 100 m, it is reported to the nearest 10 m. Making these measurements is achieved by reference to objects at specified distances from a vantage point. This is feasible on land, but at sea there are no objects conveniently located at fixed distances from ships! The scale used for marine visibility is therefore much more 'basic' and is logged only once each day. At night, reports are based on unfocused lights of moderate intensity at known distances, and where appropriate, from the silhouettes of mountains or hills against the sky.

Past and Present Weather

To meteorologists, 'weather' means more than 'wet and windy': it's something much more specific, being related to prevailing conditions at the time of the observation. There is always 'something to report'. Present weather types are allocated a distinct two digit number from 00 to 99: the larger the number, the more 'significant' the weather. Past weather is also important: observers choose two broad weather types from a list of ten. The period during which the factor is reported depends on the time of the observation: for example, reports made at 0000, 0600, 1200 and 1800 GMT/UTC relate to the previous six hours.

The observations of the weather form the basis of the station plot that appears on surface weather maps. Because all the weather stations and services around the world use the same system of codes and symbols, once the codes and symbols are learned, anyone can read them.

SYNOPTIC METEOROLOGY

By international agreement and with instructions, regulations and standard symbols that are intelligible to all users across the globe (set by the World Meteorological Organisation), weather stations across the world simultaneously observe and report the local weather four times a day, every day. The internationally agreed symbols used by meteorologists can be seen in the charts below.

Weather station observations across the world take place at 0000 GMT (Greenwich Mean Time) – also called UTC (Universal Time Co-ordinated) – and at six-hour intervals thereafter, although most stations also make observations at the intermediate 'synoptic times' of 0300, 0900, 1500, and 2100 hours. Each country's national weather service then prepares what are known as synoptic weather maps. Synoptic means presenting a broad overview of existing conditions. Unlike road maps – or even nautical or astronomer's 'star maps' – weather maps cannot be

RIGHT: *Inernationally recognised weather symbols.*

'fixed': while the stars and seas remain pretty well unchanged over our lifetimes, the weather changes daily, or even by the hour!

Synoptic meteorology began in the mid-19th century when European countries began exchanging information on the weather – especially information about storms. This began when weather observers realised that variations in the weather were directly related to barometric fluctuation and they began drawing charts marked with lines called isobars which linked together areas of equal pressure. In 1857, the Dutch meteorologist Christoph Buys Ballot cast new light on the dynamics of air pressure and its influence on the weather by showing that wind flow always follows isobars. The creating of daily weather maps developed in the 1860s and, today, meteorological services across the world generate surface and high-altitude maps based on information gathered by modern meteorological technology including radar and satellites.

The worldwide observations taken at 0000 GMT and 1200 GMT are the data that are most often used to provide the 'ingredients' for computer models that form the basis of weather predictions. One of the world's leading weather forecast

LEFT: *NEXRAD Doppler Radar Station at National Weather Service in San Angelo, Texas*

ABOVE: *Weather computers now make tracking and forecasting a little easier for meteorologists.*

centres is the European Centre for Medium-Range Weather Forecasts (ECMWF) based in Reading, in Berkshire, England. On a typical day, the ECMWF collects data from:

- *Over 6,000 weather buoys, both moored, and drifting across the seas and oceans.*

- *Over 10,000 surface observations made mainly from commercial seagoing vessels.*

- *Over 41,000 weather reports from 'traditional' weather stations on land including automated sites.*

- *Over 500 observations from automated instruments that sense wind direction and strength in the lower troposphere.*

- *Over 700 observations made by optically tracking the drift of small weather balloons to deduce the winds in the lower troposphere.*

- *Over 1,000 reports representing information from radiosonde ascents made around the world and which include data on temperature, humidity and wind direction and wind speed up to height of about 30 km (20 miles).*

- *Over 8,000 SATOBs – estimates of wind speeds and directions gained by tracking clouds from geosynchronymous satellites such as Metostat. Also known as 'geostationary' satellites, geosynchronymous satellites are so-called because they complete one orbit of the Earth (above the Equator at a height of 35,900 km/22,000 miles) in the same time that it takes for the Earth to complete one revolution (i.e. 24 hours). They therefore remain almost perfectly stationary above a fixed point and have a field covering about one-third of the Earth's surface. By international agreement, when a satellite comes to the end of its life, it is moved into a higher orbit to allow its previous site to be used.*

- *Over 32,000 aerial observations made from commercial aircraft, either automatically sensed and transmitted, or sent directly from the flight deck by the crew.*

- *Over 100,000 readings which are vertical profiles of temperature and humidity, sensed by NOAA (National Oceanic and Atmospheric Administration – the USA's national meteorological and oceanic service) polar-orbiting satellites.*

- *Over 1 million ERS-2 (European Remote Sensing Satellite) observations which give estimates of sea-surface winds.*

BELOW: ***Radar map of storm approaching California.***

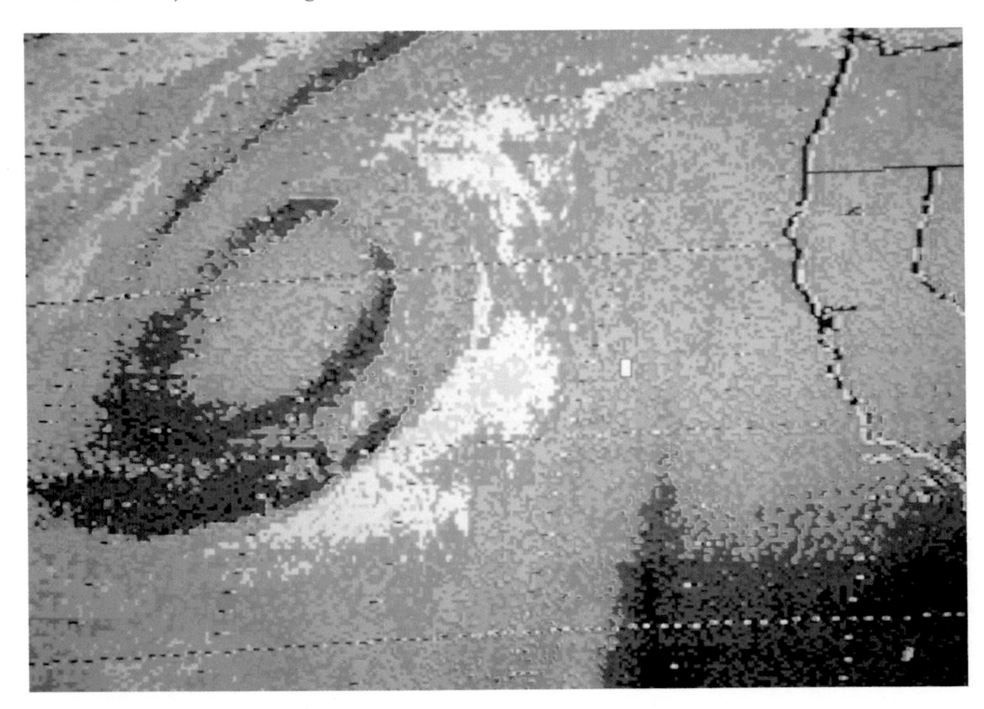

NUMERICAL WEATHER PREDICTION (NWP)

Synoptic weather forecasting remains the basis of all weather forecasting, including what is called Numerical Weather Prediction (NWP). This type of meteorology dates from around the time of World War I. Using the weather observation network in place at the time, English meteorologist Lewis Fry Richardson devised a grid network to make numerical calculations. The globe was divided into a chequer board of black and white squares, and at the centre of each dark square pressure was measured. In the centre of each light square, velocity was plotted. The problem that faced Richardson was that the enormous mountain of calculations required a great deal of time to process: this was a task ideally suited to computers, but at the time, none existed. Richard also made a vital mistake in figuring his estimated values based on observations made at six-hourly intervals: today, for a forecast even to be marginally accurate, meteorologists agree that measurements need to be taken every 30 minutes at the most! Richardson published his research in 1922 and estimated that it would take a huge weather forecasting 'army' of some 64,000 people working around the clock just to calculate the figures for a weather prediction. It's not surprising then that numerical forecasting was abandoned.

In 1939, however, Swedish meteorologist Carl-Gustav Rossby revisited Richardson's methods and by the end of the World War II, sufficient advances had been made in 'calculation technology' to make NWP more possible. The first major problem for the new computer technology that emerged in the post-war period was to 'solve' the problem of weather prediction.

From its orbiting perch high above Earth, the SeaWinds scatterometer on Midori 2 ('midori' is Japanese for the color green, symbolizing the environment) will provide the world's most accurate, highest resolution and broadest geographic coverage of ocean wind

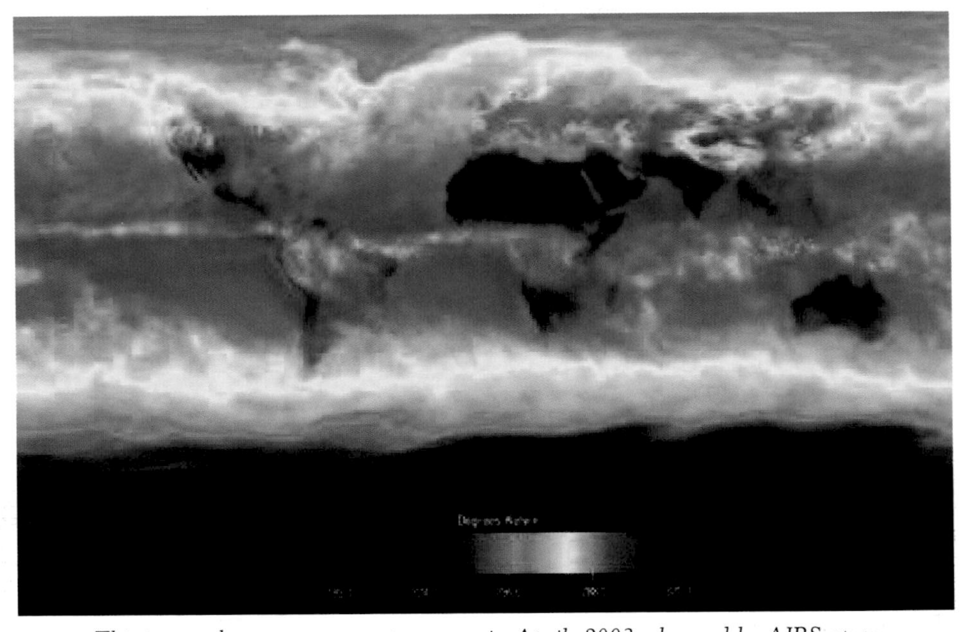

ABOVE: *This image shows average temperatures in April, 2003, observed by AIRS at an infrared wavelength that senses either the Earth's surface or any intervening cloud. Similar to a photograph of the planet taken with the camera shutter held open for a month, stationary features are captured while those obscured by moving clouds are blurred. Many continental features stand out boldly, such as our planet's vast deserts, and India, now at the end of its long, clear dry season. Also obvious are the high, cold Tibetan plateau to the north of India, and the mountains of North America. The band of yellow encircling the planet's equator is the Intertropical Convergence Zone (ITCZ), a region of persistent thunderstorms and associated high, cold clouds. The ITCZ merges with the monsoon systems of Africa and South America. Higher latitudes are increasingly obscured by clouds, though some features like the Great Lakes, the British Isles and Korea are apparent. The highest latitudes of Europe and Eurasia are completely obscured by clouds, while Antarctica stands out cold and clear at the bottom of the image.*

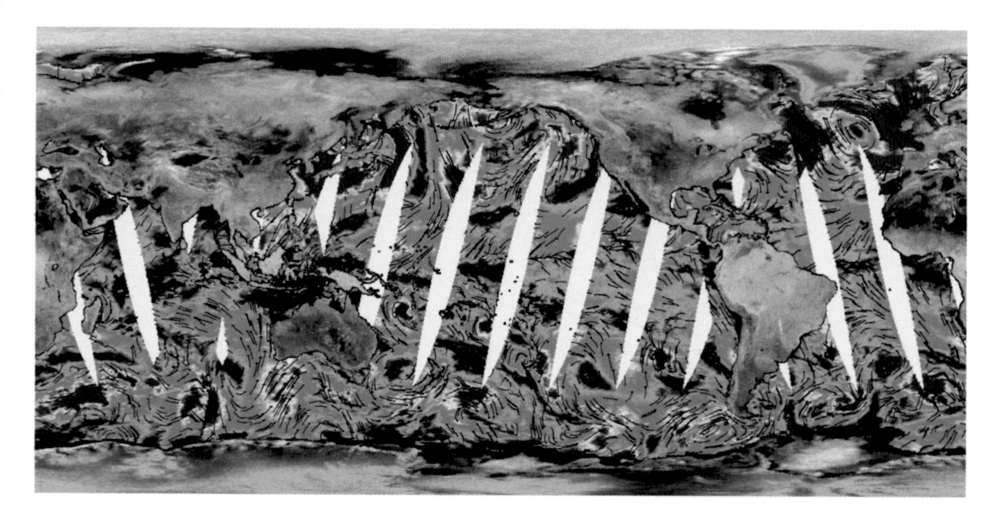

speed and direction, sea ice extent and properties of Earth's land surfaces. It will complement and eventually replace an identical instrument orbiting since June 1999 on NASA's Quick Scatterometer (QuikScat) satellite. Its three- to five-year mission will augment a long-term ocean surface wind data series that began in 1996 with launch of the NASA Scatterometer on Japan's first Adeos spacecraft (now named Midori 2).

Climatologists, meteorologists and oceanographers will soon routinely use data from SeaWinds on Midori 2 to understand and predict severe weather patterns, climate change and global weather abnormalities like El Niño. The data are expected to improve global and regional weather forecasts, ship routing and marine hazard avoidance, measurements of sea ice extent and the tracking of icebergs, among other uses.

ABOVE: *One of NASA's Earth-observing instruments, the SeaWinds scatterometer aboard Japan's Advanced Earth Observing Satellite (Adeos) 2 – now renamed Midori 2 – has successfully transmitted its first radar data to our home planet, generating its first high-quality images.*

The next major step forward came in 1950 when mathematician John von Neumann, at the Institute for Advanced Study at Princeton University in USA, published the results of his research into computerised numerical forecasting. Neumann used an enormous ENIAC digital computer which measured 25 m wide, 37 m deep and was 3 m high, and weighed in at about 30 tonnes! Despite its size – and energy consumption – this state-of-the-art machine was far less powerful than one of today's home computers!

ABOVE: *Dividing by map grid.*

Nevertheless, Neumann (or at least the computer!) was able to generate simple model-driven forecasts and the first hurdle in the race to make weather predictions using computers had been jumped. With advances in computer technology in the 1960s, NWP soon became the basis for weather forecasting throughout the world – but it is still necessary to gather huge quantities of data.

DIVIDING UP THE PLANET

To make the process of calculating data systematic, meteorologists use a variation of Richardson's' chequer board. The surface of the Earth is divided into a mesh of equally spaced locations (or grid points). Above the surface of the Earth, the atmosphere is also divided into vertical levels to create three dimensional 'stacks' of box-like compartments. Once the 3D grid has been laid out, the computer assigns various values of atmospheric wind, pressure, temperature and humidity to each grid point. Raw data supplied from surface- and upper-air observations are fed into the computer which then statistically manipulates the information to produce the values for each 'box' or grid point. The accuracy of the readings of the initial conditions – as well as the speed they are sent and received (because, as the weather changes, the recorded data becomes obsolete) – determines how accurate a numerically calculated forecast will be.

The forecasts prepared by ECMWF use a global model which

has the grid points set out horizontally across the Earth's surface at around 60-km (36-mile) intervals. Each grid point has a 'stack' of 31 'boxes' above it stretching into the atmosphere (this is to be increased shortly to 50). This model is used each day to prepare the 3-day and 10-day global forecasts, as well as global wave forecasts (although for the smaller, yet very crowded, European waters, a grid spaced at around 30 km (18 miles) is used).

Ensemble forecasts

In addition, the ECMWF uses a horizontal grid with the grid points spaced at around 120 km (120 miles) to produce 'ensemble' global forecasts. Ensemble forecasting is the method by which meteorologists estimate the likelihood that their particular forecasts will be correct – they're predicting the accuracy of their predictions!

Ensemble forecasting is done by running computer simulations a number of times, each with the initial weather conditions slightly changed. The ECMWF currently calculates 51 different versions for its global ensemble forecast. If the forecasts remain similar 7-10 days ahead, it can be assumed that the weather is 'quasi-stable' and that the predictions will be reasonably accurate. If the forecasts 'deviate' after a few days, it's more likely that weather conditions will break down suddenly and unpredictably. The solutions to the data equations are generated as images and graphics, primarily maps, of anticipated rain, wind and temperature patterns which can be put on to a computer screen, 'run together' in a sequence, and then 'fast-forwarded' to 'see' what the weather might be in the near future based on the inputted data.

The global picture of possible atmospheric conditions that is made possible by NWP models make them useful to a range of people and industries: for example, shipping companies need to know what surface winds might be doing; airlines need to know what's happening in the high-altitude winds; governments need to warn their citizens about hazardous weather, such as floods following predictions of rainfall; utilities companies supplying gas, electricity and water to peoples' homes need to know about temperature patterns; farmers and growers need to have information about anticipated weather so they can plan their crop- and stock-maintenance. Miscalculations can mean ships forced to stay in harbours, chaos at airports as flights are delayed and diverted, power and water shortages, and empty supermarket shelves. While in developed nations, normal services are soon restored, in developing nations, unpredicted weather can mean famine, disease and death.

100% ACCURATE PREDICTIONS? NOT YET!

Powerful computers, orbiting satellites, and automated weather stations which send data electronically direct to weather centres for analysis are still not enough to ascertain prevailing weather (weather conditions at the time) with absolute accuracy. The problems: first, weather observation stations are not evenly distributed over the Earth's surface, and while coverage is good for the northern hemisphere, it is less widespread in the southern hemisphere, especially in the less densely populated regions, and is pretty inadequate over the vast areas of oceans! This is exacerbated by a second problem: there is (so far!) no mathematical equation for calculating temperature, air pressure and humidity as variables of distance – knowing what the air pressure is in London or Rome cannot give us information about what the air pressure is in Paris or Madrid. Consequently, all the supercomputers at the weather centres can do is try to 'fill in the gaps' and provide rough estimates by assigning the 'most probable' values to each grid intersection based on the analysis of data from the weather stations that are the closest to those intersections. Using estimated values as their starting points, the resulting mathematical calculations can only give a possible 'glimpse' or insight, into the future weather and not a complete 'window' on the weather.

THE BUTTERFLY EFFECT

THE 'BUTTERFLY EFFECT' is a term derived from the research of Edward Lorenz, an American theoretical meteorologist at Massachusetts Institute of Technology. In 1972, he wrote a paper titled: 'Predictability: Does the Flap of a Butterfly's Wings in Brazil Set off a Tornado in Texas?' which is commonly 'interpreted' to mean that tiny disturbances in the atmosphere may become amplified, giving rise to much larger – even catastrophic – effects. Lorenz showed that the dynamics of the atmosphere are dependent on initial conditions and, therefore, that they varied immensely depending on the most minor differences in the weather. When he computed a set of starting numbers, first to six, then to three decimal places, widely divergent weather patterns were predicted. In short, a tiny atmospheric event – say, the flapping of a butterfly's wings – could cascade through the 'system' and have dramatic consequences elsewhere in the world. Lorenz's research into what has become known as Chaos Theory, has led many meteorologists to conclude that, while no such 'butterfly effect' exists, there are limits to predictability: no matter what methods are used, forecasts beyond two weeks are impossible because the weather remains unpredictable owing to our incomplete knowledge of the physics of the atmosphere.

BELOW: *Despite major advances, we still cannot predict the weather with 100% accuracy.*

GLOSSARY

Glossary

Aerosol *Very fine liquid droplets or microscopic solid particles suspended and dispersed uniformly in the atmosphere.*

Air Mass *A large body of air whose temperature and humidity are homogenous across an area several hundred kilometres wide.*

Air Pressure *The force exerted by a hypothetical column of air extending from the surface of the Earth to the outer limit of the atmosphere.*

Anticyclone *A region of high atmospheric pressure caused when air descends from the upper troposphere in its core, causing it to become the source of out flowing air.*

Atmosphere *The gaseous envelope surrounding the Earth and which consists of various layers.*

Aurora *[Latin: 'dawn'] A luminous phenomena which occurs in the upper atmosphere and often visible from high latitudes.*

Barometer *An instrument used to determine atmospheric pressure.*

Blizzard *A very cold, strong wind (Force 7or above) laden with large quantities of blowing snow.*

Contrail *[Condensation Trail] A trail of water droplets or ice crystals produced by condensation or freezing of the water vapour present in aircraft exhaust gases.*

Corona *[Latin: 'crown'] 1) An optical phenomena consisting of one or more sets of coloured rings that surround the Sun or Moon. 2) The incredibly hot outer atmosphere of the Sun.*

Cyclone *1) The generic name for a weather centred on a low-pressure system where the winds flow inwards. 2) Local name for a Tropical Cyclone in the Indian and western South Pacific Oceans.*

Dew *Water droplets that have condensed on grass, leaves and other objects close to the ground surface as a result of overnight cooling.*

Dew bow *A coloured bow - similar to a rainbow - seen on dew-covered surfaces.*

Diffraction *The dispersion, deviation or 'bending' of rays of light from their normally straight path as they encounter the edge of a physical barrier such as a water droplet.*

Distrail *[Dissipation Trail] A clear lane in a cloud produced by an aircraft and therefore, the opposite of a contrail.*

Dry Air *In physical meteorology, dry air is air that*

contains no water vapour whatsoever.

Fata Morgana *[Italian: 'Morgan the Fairy'] A form of superior mirage. Fata Morgana are named after Morgan, the half-sister of King Arthur, who was able to create such illusions.*

Ferrel cell *A mid-latitude circulation cell described by the American meteorologist William Ferrel (1817-1891).*

Fog *A visible suspension of water droplets in the atmosphere near the Earth's surface and defined as reducing visibility to less than 1 km.*

Fog bow *A white arc with a radius of approximately 42 degrees centred on the antisolar point (i.e. opposite the sun). The water droplets are so small that diffraction (see above) broadens the bands of colour so they overlap and appear white.*

Gale *A wind of Force 8 which is measured at the standard anemometer (a device for measuring wind speed) height of 10 meters (approx. 30 ft) above ground.*

Glaze *Also known as 'black ice', glaze is a transparent layer of ice that forms when drizzle, fog (see above) or raindrops freeze on contact with a cold surface.*

Hail *Solid precipitation (see below) in the form of balls or pellets of ice, generally taken to be 5 mm (approx. 1/2 inch) in diameter or larger.*

Halo *One of many different rings, arcs and points of light that can be seen around the Sun or Moon. Caused by refraction and reflection by ice crystals either in high altitude clouds or in smaller crystals suspended in the atmosphere.*

Haze *Atmospheric 'obscuration' caused by dry particles small enough to remain suspended for long periods and which give a pearly quality to daylight.*

Hoar frost *Soft ice crystals that form on plants or other objects that have cooled to below 0 degrees.*

Hurricane *The term used for a tropical cyclone that occurs in the Atlantic, Caribbean and Eastern Pacific Ocean and whose surface wind is measured as Force 12 on the Beaufort Scale.*

Intertropical Convergence Zone *(ITCZ) The region over the tropical oceans where the trade winds from the northern and southern hemispheres converge. Marked by a more or less continuous band of cumuliform clouds.*

Jet Stream *A narrow current of high-speed winds, thousands of kilometres long, hundred of kilometres wide and a few kilometres deep, that occur in the upper troposphere and lower stratosphere.*

Mesosphere *The atmospheric 'layer' lying between stratosphere and the overlying thermosphere, at an altitude of c. 50 and 83-103 km.*

Mirage *An optical phenomenon caused by refraction in the lowest layers of the Earth's atmosphere that affects the range of distant objects.*

Mist *A suspension of small water droplets that slightly obscure visibility. Mist is reported when visibility exceeds 1 km: visibility below that distance and the obscuration would be classed as fog*

Monsoon *[from Arabic mausim: 'season'] A wind that is persistent in one direction throughout a season, and which changes direction from one season to another. Also the rainy season that occurs with onset of the south-west monsoon wind in Asia.*

Ozone Layer *The layer in the atmosphere between 10 and 50 km that contains most atmospheric ozone which absorbs most of the energy from ultraviolet radiation from the Sun and thus acts as a 'protective shield' for life on our planet.*

Precipitation *Water, either in liquid or solid form, derived from the atmosphere (see above) which falls to the surface of the Earth.*

Rainbow *An optical phenomena that appears in the form of one or more arcs of spectral colours (red, orange, yellow, green, blue, indigo and violet) when sunlight is both reflected and refracted by falling raindrops.*

Reflection *The process by which some or all radiation (including light) is returned in the general direction of the source. The exact amount of reflection is dependent on the wavelength of the radiation, the nature of the reflecting surface and the angle of incidence.*

Refraction *The deviation of light (or other electromagnetic radiation) from its original path. Refraction is usually accompanied by dispersion.*

Rime *A deposit of ice that forms as rough crystals through the freezing of super cooled water droplets from fog on contact with solid surfaces.*

Sleet *In Britain, sleet is mixture of rain or drizzle with melting snow. In the US (but not Canada) sleet refers to ice pellets.*

Stratosphere *The layer in the atmosphere (see above) that lies above the troposphere (see below) and below the mesosphere (see above).*

Super cell storm *An extremely violent and persistent thunderstorm characterised by the formation of extremely large, rotating updraughts of air that extend high into cumulonimbus clouds (which may lie at about 8- 15 km).*

Thermosphere *The outermost layer of the Earth's atmosphere (see above) which lies above the mesospause (at c. 86-100 km) and extends into interplanetary space.*

Tornado *A violent, rapidly rotating column of air that extends downwards from the base of cumulonimbus cloud and reaches the ground. The intensity of tornadoes is measured on the Fujita-Pearson Scale or the TORRO scale.*

Troposphere *[Greek: 'sphere of change'] The troposphere is the lowermost layer of the Earth's atmosphere (see above) extending from the surface to about 7 km altitude near the poles, and 14-18 km at the Equator.*

Typhoon *[From Chinese tai fun: 'great wind'] The name given to tropical cyclones (see above) in the western North Pacific.*

Waterspouts *A rapidly rotating column of air that occurs over a lake, river or sea. Waterspouts tend to dissipate as soon as they cross onto dry land.*

BIBLIOGRAPHY

Be Your Own Weather Expert K. Baker, Merlin, 1995

Atmosphere, Weather and Climate R.G Barry & R.J Chorley, Routledge, 1995

Weather: The Ultimate Guide to the Elements W.J Burroughs, Harper Collins, 1996

How The Weather Works R. Chaboud, Thames and Hudson, 1996

Collins Gem Weather Photoguide S. Dunlop, Harper Collins, 1996

Weather: Get to know the Natural World S. Dunlop, Collins, 2004

How to Identify the Weather S. Dunlop, Collins, 2002

Oxford Dictionary of Weather S. Dunlop, Oxford University Press, 2001

Weatherwise P. Eden, MacMillan, 1995

Daily Telegraph Book of the Weather P. Eden, Continuum, 2003

Weather Facts D. File, Oxford University Press, 1996

Meteorological Office Cloud Types for Observers, Stationery Office, 1982

Philip's Guide to Weather R. Reynolds, Philip's, 2000

Heaven's Breath: A Natural History of the Wind L. Watson, 1984

Weather: The Ultimate Guide to the Elements R. Whittaker (ed.) Collins, 1996

The Weather Book J. Williams, USA Today, 1992

Websites:

BBC Weather: www.bbc.co.uk/weather
UK Weather Information Site: www.weather.org.uk
CNN Weather: www.cnn.com/WEATHER/index.html
The Weather Channel: www.weather.com
NOAA (National Oceanic & Atmospheric Administration): www.noaa.gov
NOAA National Climata Data Centre: www.ncda.noaa.gov
National Severe Storms Laboratory: www.nsl.noaa.gov
National Hurrican Center, USA: www.nhc.noaa.gov
ECMWF (European Centre for Medium Range Weather Forecasting): www.ecmwf.int
Meteorological Office, UK: www.metoffice.com
National Weather Service, USA: www.nws.noaa.gov
World Meteorological Organisation: www.wmo.ch

Societies:
American Meteorological Society: www.ametsoc.org/AMS
Canadian Meteorological and Oceanographic Society: www.meds.dfo.ca/cmos
European Meteorological Society: www.emetsoc.org
National Weatherr Association, USA: www.nwas.org
Royal Meteorological Society: www.royal-met-soc.org.uk
TORRO (Tornado & Storm Research) www.torro.org.uk

Satellite Images:
Eumetstat: www.eumetstat.de/
Metostat Geostationary Satellite Images
Remote Imaging Group: www.rig.org.uk
Polar-Orbiter Images courtesy of University of Strasbourg: www.-grtr.u-strasbg.fr

INDEX

CREDITS

Thanks once again to Vic Swift at the British Library, the staff at the Science Museum, London, and Cap'n Bob on the 'Moonbeam' for explaining the true intricacies of the shipping forecast.

Picture credits:

Images pp 6, 70, 126, 128, 136, 212b © Stockbyte.
Images pp 9, 10b, 11-12, 14, 16-18, 20, 24, 32, 36, 39, 41-42, 59, 67, 79, 84, 90-91, 96, 99, 104, 123t, 123m, 143, 196-197, 216 © Getty Images.
Images pp 15, 23, 27-29, 31, 33, 38, 52b, 54, 60, 82, 102, 110-111, 116, 118, 123b, 124-125, 130, 133-134, 156, 171, 201, 204, 206t, 209, 212t, 217, 220, 241-242, 246 © Corbis.
Images pp 35, 44-45, 51, 72, 75-76, 80, 103, 121, 131-132, 144, 225, 244-245 Courtesy NASA/JPL-Caltech.

Other images ©:
P10: William McKelvie. P15: Bill Gruber. pp21, 48t: Arlene Gee. p34: Leah-Anne Thompson. pp40, 82: Ravi Tahilramani. pp47, 207: Lars Lentz. p48b: John Archer. p49: Allen Johnson. p56: Yanik Chauvin. p66: Sergey Kashkin. p68: Kim Freitas. pp70t, 194: Dane Wertzfeld. p78: Maureen Perez. p86: Chris Auld. p87: Emma Holmwood. p88: Matthew Scherf. p92: Renee Lee. p95: David Freund. p97t: Julie Askins. p97: Kenneth C. Zirkel. p98: Donald Gruener. p100: Mark Bond. p106: Sonja Foos. p108: Gregor Erdmann. p109: David Rose. p113: Jennifer Duncan. P119: Andy Lim. p138: Jennifer Zimmerman. p140: Paul Senyszyn. p146: Fabio Frosio. p147: Stan Rohrer. p148: Tony Colter. p153: Richard Leloux. p155: Angela Sorrentino. pp156, 190b, 224t: David Gilder. p158: Carsten Madsen. p159: Rob Stegmann. p161: Nathan Goodman. p164: David Elfstrom. p165: Ana Abejon. p166: Sam Lee. p168t: Galina Dreyzina. p168b: Bill Grove. p169: Simon Moran. p170: John DeFeo. p172: Randolph Pamphrey. p173: Nicola Stratford. p174: Patrick Abanathy. p175: Dan Brandenberg. p176: Sam Aronov. p178: Fred van Engelen. p179: Falk Kienas. p180: Yohan Juliardi. p183: Ben Kopilow. p186: Brendan O'Boyle. p187: Shaun Lowe. p189: Debbie Martin. p190t: Tomaz Levstek. p191: Jean-Yves Benedeyt. p192: Timothy Babasade. p199: Cameron Schaus. p200: Amy Goodchild. p202: Karen Grotzinger. p203: Justin Horrocks. 206b: Luis Carlos Torres. p210: Dennis Glory. p213b: Guy Erwood. p211: Bertrand Collet. p213t: John Rattle. p214: Martin Pietak. p219: Jacob Carroll. p224b: Dennis Cox. p227: Victor Kapas. p228: Jeff Comfort. p229: Rob vanNostrand. p230: Israel Talby. p234: Scott Cressman. pp235, 236: Terry Healy. p239: Roberta Casaliggi. p240: Jon Sharp. p249: James Goldsworthy.